N. F. Warren

Compliments

C. O. Brawner

STABILITY IN OPEN PIT MINING

Proceedings of the
First International Conference on Stability in Open Pit Mining
Vancouver, B.C., Canada, November 23-25, 1970

Edited by

C. O. BRAWNER
and
V. MILLIGAN

Sponsored by

Center for Continuing Education, University of British Columbia
Engineering Institute of Canada (B.C. Section)
Canadian Institute of Mining and Metallurgy (B.C. Section)

Society of Mining Engineers
of
The American Institute of Mining, Metallurgical, and Petroleum Engineers, Inc.

New York, 1971

Printed in the United States of America
by Port City Press, Inc., Baltimore, Maryland

Library of Congress Catalog Card Number 75-157732

CONFERENCE ORGANIZING COMMITTEE

C. O. BRAWNER	*Vice President* *Golder, Brawner and Associates Ltd.,* *Vancouver, B.C., Canada*
L. G. R. CROUCH	*Professor of Mineral Engineering* *University of British Columbia* *Vancouver, B.C., Canada*
M. K. DOUGLASS	*Supervisor of Lands and Services* *Brameda Resources Ltd.,* *Vancouver, B.C., Canada*
H. G. EWANCHUK	*Vice President* *Bethlehem Copper Corp. Ltd.,* *Vancouver, B.C., Canada*
G. T. MATTHEWS	*Administrator* *Center for Continuing Education* *University of British Columbia,* *Vancouver, B.C., Canada*

CONTENTS

Chapter 1

INTRODUCTION

by
C. O. Brawner

Principal, Golder, Brawner and Associates Ltd., Vancouver.

The advent of larger drilling, excavation, and milling equipment is resulting in a tremendous increase in the scale and annual tonnage of open pit mining. At the present time, open pits up to 3000 ft. deep, and tailings dams and waste dumps up to 600 ft. high are being planned.

One of the major problems that projects of this magnitude create is stability. Stability at open pit mining developments must be assessed for tailings dams, waste dumps, and open pit slopes in overburden soil and in rock.

Soil mechanics principles are used to assess the stability for the first three. This science was developed by Dr. Karl Terzaghi in the 1920's, and since that time the theories and principles have been tested and evaluated. This experience has resulted in the development of sound theoretical concepts, workable sampling and testing techniques, and accurate methods of analysis. Typical problems that can be analyzed include:

— bearing capacity of soil under dams and waste dumps.
— amount of seepage under and through dams.
— stability of slopes and determination of safe slope angles for tailings dams, waste dumps, and open pits.
— influence of earthquakes on stability.
— influence and cost of different excavation and construction techniques.

To assist in comparing engineering experience, an international soil classification system has been developed and used for several decades. In addition, many excellent text books are available outlining soil mechanics principles and techniques.

A high degree of technology exists in soil mechanics. This, the mining engineer cannot afford to ignore.

If soil mechanics principles had been applied, many major failures involving tailings, tailings dams, or waste dumps, such as those at Aberfan, Mufulira, Luanshya, and El Cobre would not have occurred.

Tailings dams are planned in Canada that will be up to 600 ft. high. Construction procedures used in the past have paid little attention to the compaction of the shell material or to the influence of earthquakes on stability. This neglect has led to failures. While failures in low dams may not be too serious, for high dams such failures could have catastrophic results.

Another major area of concern is the stability of rock slopes. From the standpoint of long term safety and economics, rock mechanics problems are frequently more serious than soil mechanics problems in open pit mining. This is partly due to the increasing depths of proposed open pit mines. As an example, on three recent projects that the author has been associated with, each degree of slope was stated to represent a potential cost, or saving, of 5 to 15 million dollars.

Rock mechanics as applied to slope stability is at about the same stage that existed in soil mechanics 30 years ago (1). Surprising as it may seem, very few significant theoretical contributions were made before 1960. Whereas the First International Soil Conference was held in 1936, the First International Rock Mechanics Conference was only held in 1966.

There are several reasons for the late development of rock mechanics in relation to surface mining.

- until the last decade, the majority of mines have been underground.
- most senior mining engineers have been trained in underground mining.
- university courses in rock mechanics have only been established in recent years.
- there has been a reluctance to spend money on rock mechanics because of an apparent lack of certainty of economic reward.

A further major problem is that rock mechanics is more complicated than soil mechanics. Some of the reasons for this are the following:

- rock materials are non-homogeneous and usually anisoptropic.
- strength parameters relating to rock masses are infinitely variable and difficult, if not impossible, to determine precisely.
- generalized models and theories of rock behaviour are complex, as are the mathematics involved.
- field conditions are extremely difficult, and often impossible to duplicate in the laboratory.
- field testing is usually complicated, time consuming, and almost always very expensive.

There have been significant advances in our knowledge of rock mechanics in the past 10 years. Unfortunately, there has been limited coordinated effort to assimilate this information and assess the advances in relation to our field experience.

The theme of this symposium is to review, in practical terms, what we actually know about rock mechanics relating to slope stability. Most of us have been away from university mathematics for many years so that a detailed theoretical treatment would be a waste of time.

An important factor to recognize is that while rock mechanics is a new science, we can make use of theory and experience from many other related fields. The following are examples:

- the theory of elasticity may be used for studies of stress distribution in rock masses.
- the techniques of triaxial and uniaxial compression, direct shear and permeability tests, as applied to soils, can also be used to test rock.
- the principle of effective stress and the mechanics of the stability analysis, developed many years ago in soil mechanics, are applicable to rock slopes.
- the measurement of water pressures and the flow of water through porous media can take advantage of the experience used in investigation for the construction of major dams. These studies date back to the 1930's.
- geologic mapping techniques to classify rock and to define the orientation and frequency of discontinuities are essential in rock mechanics studies.
- the principles and techniques of stabilization of landslides in soil are usually equally effective for many types of slides in rock.

The lectures also outline new developments in rock mechanics. The advent of the computer and the finite element technique have made it possible to incorporate the third dimension in stability studies. However, these new techniques are not a cure-all. When used with discretion, in the hands of experienced engineers, they are a powerful tool. In the hands of the inexperienced, they may only provide a quicker way of getting the wrong answer.

Many fallacies regarding stability have developed over the years. Part of the challenge to the authors will be to emphasize and correct these. Typical of these corrections are:

- water does not generally act as a lubricant in slides.
- reduction in water pressure is a more important factor in stability than is the amount of water that is intercepted.
- the influence of blasting on stability is far more important than most mining engineers recognize.
- placement of tailings by hydraulic means does not provide a high degree of density.
- the length of time that the seismic vibrations last during an earthquake can have a major influence on stability.

If an open pit operation does not have some evidence of instability, money is being wasted. However, if potential instability exists and is unsuspected, lives may be lost. Controlled stability is good mine management. Accordingly, operating mine engineers should have a knowledge of the general techniques of stabilization. This should also include a knowledge of the instrumentation used to monitor slope movement.

The late Dr. R.A.L. Black, former Head of the Dept. of Mines, Royal School of Mines, London, and Dr. E. Hoek, Head of the Rock Mechanics Project, made some very pertinent comments at the 1967 Conference in Golden, Colorado (2).

> "Rock mechanics is no longer only an esoteric study for academic initiates: It is the essential basis of practical mine design."
>
> "A body of really useful knowledge in rock slope stability cannot be developed unless the more academic studies are developed and validated in practice in the field. There are no short cuts, no cheap solutions and no substitutes for a proper and complete appreciation of the basic problems involved. The nature of the forces which are at work when ancient equilibrium is disturbed by the creation of mining excavations must be understood before real progress in the application of rock mechanics to mine design can be made."
>
> "The penalty for the practical engineer for attempting to defy the known laws of strata behaviour is relentlessly exacted in hard economic terms and sometimes in blood."

We must apply the knowledge that we already have. We must build the knowledge up from carefully validated experience. The work must be coordinated between research establishments, such as universities, and the mining companies concerned.

We have developed analysis and design techniques that are far beyond our capability to determine the necessary strength and boundary parameters. We are desperately in need of case studies and analysis of field failures to test theoretical concepts. Only then will real practical progress be made.

Apart from the economic advantages of applying rock mechanics principles correctly, there is the certain benefit of better and safer working conditions.

To make or to save money, we must be prepared to spend it in commensurate amounts. There are no ready made solutions in the field of mining rock mechanics, but given time and reasonable financial aid and other support from industry, solutions can be found.

REFERENCES

1. Golder, H.Q., "The Practice of Sampling Soils and Rocks". A.S.T.M. Symposium on Sampling Soil and Rock, Toronto. 1970

2. Black, R.A.L. and Hoek, E., "Status of Rock Mechanics as Applied to Mining". Ninth Symposium on Rock Mechanics, Golden, Colorado. 1967.

Chapter 2

THE ROLE OF SLOPE STABILITY IN THE ECONOMICS, DESIGN AND OPERATION OF OPEN PIT MINES

by

Richard M. Stewart
Director Mining Research, The Anaconda Company

and

Bruce A. Kennedy
Senior Mining Research Engineer, The Anaconda Company

ABSTRACT

Slope stability has now become a practical tool in the design and operation of open pits. The selection of slope angles is a critical decision that can have far-reaching effects on the accumulated cash flow and the overall profitability of a mining project. Slope failures can now be dealt with in a number of ways. Two methods — the total removal and the partial unloading of a failing zone — have been practiced for many years. However, with the refinement of monitoring techniques, open pits can now learn to live with a slide and even turn a potential loss into a gain. There are various techniques, such as presplit blasting, drainage systems, and artificial stabilization, which can be applied immediately in an existing open pit to improve the stability of the pit slopes. The technology of slope stability will play an important part in the future mining of low grade, previously marginal orebodies, which will have to be exploited if the mineral needs of the world are to be met.

INTRODUCTION

Rock can be defined as an engineering material. Just as the civil engineer works with concrete as an engineering material, the mechanical engineer with steel and other metals, the mining engineer has to consider the engineering of structures in rock. Unfortunately, close tolerance design of rock structures is extremely difficult because, unlike metal or concrete, rock is a highly complex material. It varies in its characteristics across very short distances, is usually highly fractured or broken, and is intermixed with other substances such as clays, quartz healings, etc. Further complications are added by the presence of water, and the effect of seismic loadings from blasting and earthquakes. However, the mining industry has to work with this material to achieve overall pit slopes that are both economic and safe (1) (2). This is the study of slope stability.

The primary reason for open pit slope stability studies is to design the slopes to optimize the economics of the pit commensurate with the capability of the rock to remain stable, and also to minimize the chance of failures and the consequent possible hazards to personnel and the operation (3) (4).

As larger earthmoving equipment becomes available and production costs are lowered, previously marginal orebodies are becoming more viable prospects (5). The increasing demand for minerals and metals makes lower cutoff grades and higher stripping ratios acceptable. Consequently, open pit mines are being planned today to go to depths and sizes that have never been considered before. The role of slope stability in open pit mining is sixfold (6) (7) (8) (9). It is involved with 1) design economics, 2) design safety, 3) operational economics 4) operational safety, 5) the economics of slope failures, and 6) the safety aspects of slope failures. The table below summarizes the main functions of slope stability in open pit mining and the potential economic savings and safety improvements:

SLOPE STABILITY

- Design
  - Economic
    - Reduction of stripping ratio
    - Reduction of incurred costs due to deferred stripping
    - Possible increase in ore reserves
  - Safety
    - Better awareness of condition of slopes
    - Design of support systems if required and economically justified
    - Water control—surface and underground
- Excavation
  - Economic
    - Reduction of damage to slopes and improved fragmentation from better blasting techniques
  - Safety
    - Safety Berms
- Failure Prediction
  - Economic
    - Reduction of losses due to failure
    - Ability to live with a failure
  - Safety
    - Prevention of hazards to personnel and equipment

THE ROLE OF SLOPE STABILITY IN THE DEVELOPMENT OF AN OPEN PIT

The primary reason for the existence of a mine is to make a profit, or a return on investment. As compared to other industrial concerns, mining is faced with a unique problem in that its capital assets — orebodies — are depleted. A mining company, in considering a single ore deposit, can adopt one of two possible approaches to mining the deposit. First, by rapidly mining the orebody it is possible to maximize the return on investment or; secondly, by using a lower capital investment and a longer life for the pit, the total profits from the pit can be maximized. From an accumulated cash flow basis the first system mentioned above gives a very high cash flow **out** in the initial years and a high cash flow **in** in later years over a shorter period of time; whereas the second system produces over a longer period of time and gives a more even cash flow over the life of the mine, with lower investment peaks.

The Basic Input Factors

To examine the role of slope stability in mine economics and mine planning, it is necessary to examine the basic economic decisions that have to be made before mining commences. The very first consideration is the determination as to whether the orebody should be mined by open pit or underground means. The current trend in the world today is toward mining at much greater depths by open pit methods. Underground mining costs are increasing, whereas open pit mining costs are decreasing, mainly due to the large capacity equipment now being used. Obviously, the controlling factor with regard to the choice of mining method will be the cost of removing waste material by open pit mining techniques to expose and remove the ore. This ratio of waste removed to recover one ton of ore is called the stripping ratio. If the stripping ratio becomes too large (in other words, the amount of waste material that has to be removed to recover one ton of ore becomes too great), a point is reached where it would be more profitable to revert to underground mining. It is important, therefore, to determine at what stripping ratio it would be more profitable to revert to underground mining. This is an economic ratio that would set the maximum allowable stripping ratio the open pit could go to and still be profitable. This ratio is calculated from the following formula:

$$\frac{\text{Cost per ton of ore by U/G methods} - \text{Cost per ton of ore by O/P Methods}}{\text{Cost of O/P stripping per ton of waste}}$$

This ratio then sets the limits as to what can be mined and under no circumstances should ore be mined if it exists in such a position that its actual stripping ratio is greater than this ratio.

A second stripping ratio, known as the economic limit stripping ratio, can then be determined and also applied to the final pit. This ratio limits the pit geometry to a certain profit margin and is calculated from the following equation:

Economic limit stripping ratio =

$$\frac{\text{Recoverable value/ton ore} - (\text{Production cost/ton ore} + \text{min. profit/ton ore})}{\text{Stripping cost/ton waste}}$$

This ratio ensures that a profit is made on the ore mined.

A third stripping ratio is the overall stripping ratio, which is the actual ratio of tons of waste to tons of ore that exists at the final pit limits. In no case should the overall ratio be larger than either the economic limit ratio or the ratio calculated for the cutoff between underground and open pit methods.

The overall stripping ratio does not, however, hold true as far as mining is concerned over individual years or increments of time. A fourth stripping ratio is now introduced and this is known as the instantaneous stripping ratio. The instantaneous stripping ratio is defined as the first derivative with respect to the distance in the direction of expansion of the overall stripping ratio, which is

$$\frac{\text{Total tons of material} - \text{Total tons of ore}}{\text{Total tons of ore}}$$

In other words, at any stage in the life of the pit the overall stripping ratio represents the number of units of waste material that has been removed to expose one unit of ore; and the instantaneous stripping ratio represents the number of additional units of waste that must be removed to expose one additional unit of ore.

Once the geological, metallurgical, and legal aspects have been determined, a feasibility study is made of the project. The various inputs and considerations that have to be examined are

Geological	Metallurgical	Legal	Engineering	Economic
Ore assays	Processing methods	Property delineation	Overall tonnages	Marketing conditions
Economic geology	Recoveries	Liability	Production rate	Market trends
	Byproduct values			Capital expenditure
		Ownership		
Structural geology		Royalties		
			Mining Plant	Interest rates
Extent of ore reserves			Mining equipment	Operating costs
Ore types			Pit limits	Profit required
Hydrology			Stripping ratios	
Topography			Bench height	
			Pit slope angles	
			Road grades and width	
			Ecological factors	

Financial Analysis Techniques

In analyzing the economics of any mining project, all possible economic analysis techniques must be used, since no one technique provides a complete answer. The six most commonly used are:

1) The Payback system. This system is based on the time required for the initial investment to generate income equal to the initial investment. The system is very simple but does not take into account the time value of money, taxation, interest, or profitability.

2) Average Return on Investment method. This method produces a percentage value of return on investment by determining the income generated by the investment and then dividing by the number of years of the life of the investment. This gives an average dollar value return on investment per year. This value is then expressed as a percentage of the initial investment. This method suffers from the fact that it does not consider the time value of money, interest, or taxation.

3) The Discounted Cash Flow Method. This method is more complex than those discussed above. Basically, the method breaks down the investment into equal time periods across its life and then determines the rate of return of the investment, taking into account a discount rate in which the present value of the cash inflow equals the present value of the cash outflow. This method takes into account profitability, taxation, the time value of money, and interest. However, it does not yield a final profitability point; it assumes that cash inflows are immediately reinvested.

4) The Hoskild Formula. This method of evaluation has been in use in the mining industry for almost a hundred years. Although this formula does recognize the time value of money, it has the disadvantage of using two rate values; namely, the safe rate and the risk rate, which if held constant will produce an erroneous answer. Realistically, the risk rate would have to increase as the safe rate decreases over the life of the mine.

5) The Net Present Value Method. This method is very similar to the Discounted Cash Flow method. Instead of determining the discount rate that will make the future net

cash flows equal to the initial outflow, a rate equivalent to the cost of capital is applied as the discount factor. This method has the ability to compare the present value of future profits and the present value of the capital investment that could be invested in a number of different ways.

6) Equivalent annual (or monthly) profit method. This is a technique for comparing profits, in terms of their gains per year (or month) at minimum rates of return.

The Effect of Slope Angles on the Economics of the Pit

The major factors that are considered in the economic analysis of a possible open pit are:

1. The current market price of the final product.
2. A forecast of the future market and market price of the final product.
3. The production rate of the mine.
4. The production cost.
5. The ore reserves.
6. The stripping ratio.

As mentioned above, the aim of the mine operators is to maximize the present value of future profits throughout the life of the mine. In order to do this, a certain amount of control on some factor is required by the mine management. Of the six factors shown above, management can do little or nothing to alter 1 and 2. Item 3, the production rate, is generally determined in the early planning stage and can only be changed, with some difficulty, once the mine is in production and with a further sizable capital investment. Item 4, the production cost, can be varied by very small degrees during the life of the mine. Factors 5 and 6 can both be altered to some extent by varying the geometry of the pit, the cutoff grade, or the slope angles.

To demonstrate the effect of the variation of the final slope angle on the overall stripping ratio, the tons of waste that have to be removed, and the tons of ore available, the hypothetical open pit and orebody shown in Fig. 1 will be examined. The orebody is in the form of a cylinder 900 m. in dia. and 150 m. thick, lying in a horizontal plane and overlain by 150 m. of overburden. Fig. 2 shows the variations that will occur in the stripping ratio, total material, total waste, and total ore as the final slopes are varied from 20° to 70° from the horizontal. The slopes are pivoted around the ore/overburden contact, and a constant density of 2½ tons per cu. m. for both ore and waste material is assumed. Representative values are as follows:

Slope Angle	Tons of Ore Available	Tons of Waste Material	Total Tons of Material	Overall Stripping Ratio
20°	37,929,077	218,051,851	255,980,928	5.75:1
30°	53,570,814	165,879,701	219,450,515	3.10:1
40°	64,573,895	141,393,918	205,967,813	2.19:1
45°	68,926,543	133,250,652	202,177,195	1.93:1
50°	72,759,782	126,638,685	199,398,467	1.74:1
60°	79,348,378	116,313,435	195,661,813	1.47:1
70°	85,042,733	108,291,160	193,333,893	1.27:1

As would be expected, the greatest change occurs at the lower slope angles. However, even at more traditional slope angles, the decrease in waste and increase in ore are still sizable. For example, steepening the final slopes from 45° to 50° shows an increase in available ore of almost 4 million tons, and a decrease in waste stripping of over 6.5 million tons. The overall

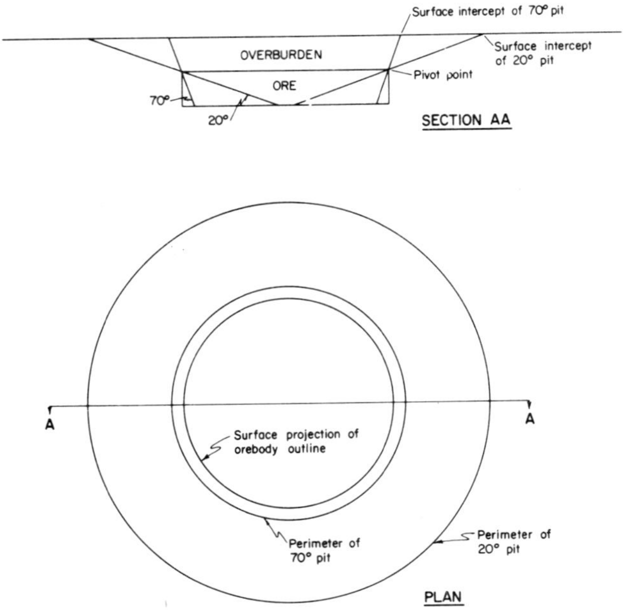

Fig. 1 —Hypothetical open pit and ore body

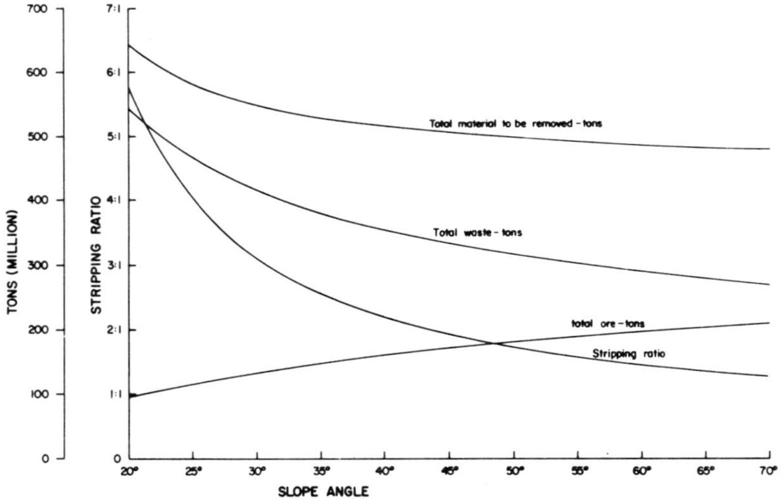

Fig. 2 —Plots of total material, total waste, total ore, and stripping ratio versus slope angle stripping ratio is decreased from 1.93:1 to 1.74:1.

Probably the most economically significant use of steep slopes is in the initial stripping of an open pit. This results in less waste being removed for the same initial ore exposure, or a given ore production, than if the flatter slope angles planned for the eventual working slopes are used. Once ore production has been initiated, the slopes can be flattened; and when final pit limits are reached, they can be steepened again. This has two economic benefits. First, in the initial stripping, the accumulated cash flow out will not peak as high as it would if flatter slope angles are used. Secondly, it is a fairly standard practice during stripping operations to set up a deferred stripping account to offset some of the waste removal against future ore to be mined. This is expensive, since interest charges have to be carried. By carrying steeper slopes, less waste material is removed and, hence, the deferred stripping account can be reduced considerably in size.

To illustrate the economic benefits to be gained by this approach, a theoretical pit will be mined using the same orebody as shown in Fig. 1. The following simplifying assumptions are made:

1) Of the gross income from the final product, the mine is credited with $8.50 for each ton of ore mined.

2) The mining cost is 25¢ per ton, irrespective of whether the material is ore or waste.

3) The mining equipment has a daily capacity of 100,000 tons, all of which is used on waste during the initial stripping. Once ore is exposed, production commences at 30,000 tons per day and remains at that level throughout the life of the mine. Stripping is continued at 70,000 tons per day.

4) An initial capital investment of $7.5 million is made, followed by a further $7.5 million spread over the first 25 months of operation.

5) The final pit slopes are at 70°.

6) No taxes or extraordinary expenses are considered.

7) The first stripping phase must expose enough ore to maintain ore production during the subsequent stripping sequence.

Two initial pits were considered (See Fig. 3). The first (Pit A) was a pit 550 m. in dia. at the surface with 60° slopes. Once the ore was completely exposed in the bottom, ore production commenced with a 45° slope in the orebody. Simultaneously, the waste slopes were flattened to a 45° working slope. Once this was completed, a second stripping sequence at 45°, which increased the surface diameter of the pit was started. The second pit (Pit B) examined was stripped with 45° slopes to give an identical ore exposure at the ore/waste contact as the 60° pit. Once production commenced, a second stripping sequence was started maintaining the 45° slopes.

The calculations revealed that ore production could be started at the end of the 23rd month with the 60° pit, and at the end of the 30th month with the 45° pit. Fig. 4 shows an Accumulated Cash Flow Chart for the first 3½ years of operation of both pits. As can be seen from this chart, the accumulated cash flow out, for the same ore exposure, by the 60° pit is about $40 million, and for the 45° pit, about $48 million. The initial capital outlay required is therefore reduced one-sixth by starting with steeper slopes. Also, the accumulated cash flow for the 60° pit crosses over to the credit side approximately 7½ months before it does for the 45° pit. Further calculations over the life of the mine showed that the present value (at an 18% interest or expected rate of return) of total future earnings for the 60° pit, at the start of stripping operations, is $162.6 million, whereas the present value of the 45° pit at the start of operations is $127.2 million. Note that these figures do not include taxes or other extraordinary expenses. Applying a capital recovery factor to the present values across the life of the mine

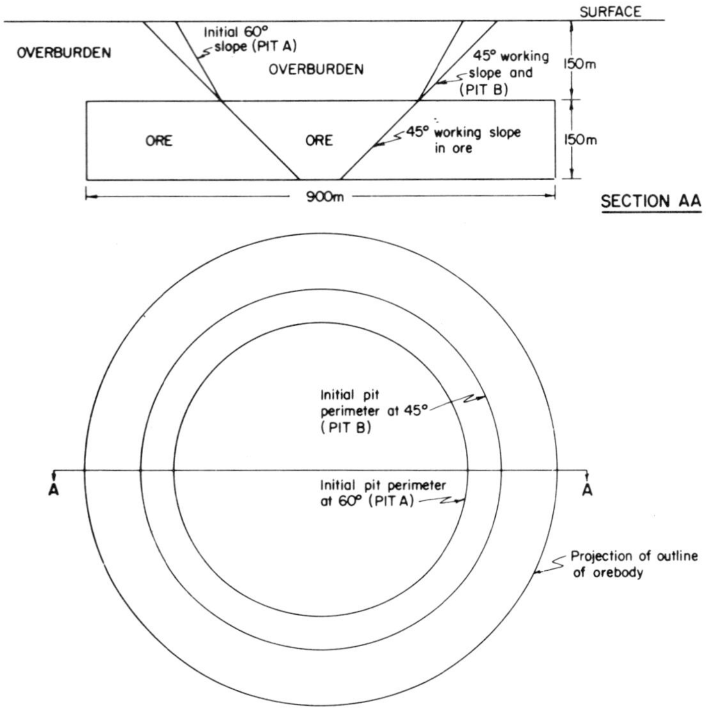

Fig. 3 —Plan and section of pit A and pit B

reveals that the equivalent earnings per month will be worth $685,000 more for the 60° pit than for the 45° pit. If all factors, such as taxes, were included, this value would more realistically be about $300,000 per month. Discounted cash flow rates of return, calculated across the life of the mine, were as follows:

Pit A — 73.4%

Pit B — 66.3%

If taxes were included, these figures would be on the order of 30% for Pit A, and 27% for Pit B. This example, even though extremely hypothetical, shows very dramatically not only the effect of using steep slope angles in the initial stripping on the initial capital outlay, but also its very drastic effect on the **overall profitability** of the project.

The economic gains of operating at steep slope angles must now be evident. However, it is also necessary to consider the factors that prohibit mining a slope at too steep an angle. The main problem is the ability to operate large mining equipment on steep slopes, and in the resulting confined spaces. The major parameters to be considered are the haulroad grade, the haulroad width, swing room for shovels, and bench widths for drills. If bench widths are maximized, the slope angles are automatically flattened and, consequently, the stripping requirements are increased. A number of trial designs of the initial pit will soon produce an optimum design in which reasonably steep slopes are attained, yet operating room is also available. It is unwise to force equipment to work in cramped space merely to attain steep slopes. Accidents and maintenance costs will increase, and productivity will decrease. It will also be necessary to leave safety berms at least on every other bench to catch loose material. In some cases, state or local law can dictate bench heights and berm widths. This will also reduce the slope angle. Consideration could be given, in the early stripping stages, to mining double benches to reduce the effect of safety berms on the overall slope angle.

Risk Analysis

Finally, owing to the natural risk that is inherent in mining investment, a careful risk analysis of the investment must be made. Because of the higher risk rate of the investment, a greater rate of return should be expected. How is the risk rate for a mining project determined? Each variable that has a probability of varying from the calculated norm has to be examined. These include the ore grade, ore tonnage, metal recoveries, production cost, capital investment necessary, etc. If the mine is to be stripped in the initial stages at steep slope angles to keep the net cash outflow peak down, a considerable increase in the risk in the early life of the pit must be expected. An exhaustive study of the structural geology from the drill holes, surface intersections, seismic mapping, plus hydrogeological work, strength testing, etc., can allow preliminary designs of the slopes to be made. A chart of Probability of Failure vs. Slope Angle can then be made (see Fig. 5) (10). From this chart, bearing in mind that the chart itself has certain limitations, a series of confidence limits can be drawn up for the slopes to be used. This data can then be analyzed with the other variables to produce a series of risk rates for the project.

Once these analyses have been made, the management of the mining company will make its decision as to whether to develop the pit or not. As can be seen from the examples above, slope stability can, and does, play a major role in the design of a profitable open pit.

THE ROLE OF SLOPE STABILITY IN OPEN PIT OPERATION

The majority of excavated rock slopes existing in open pits today were mined with no

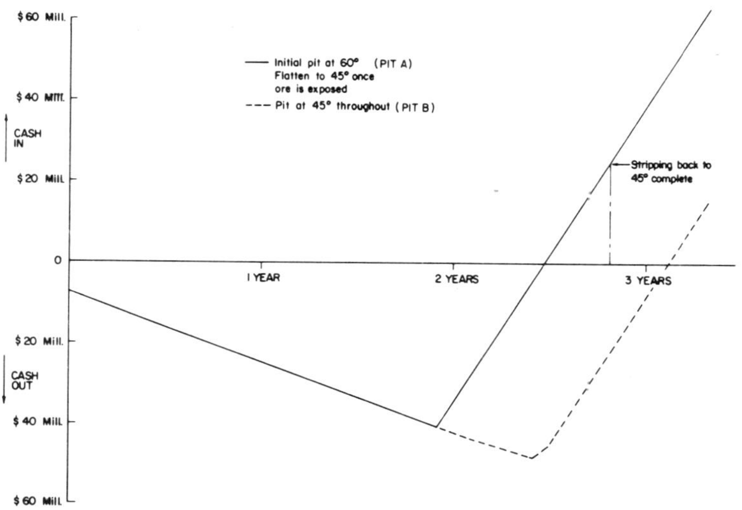

Fig. 4 —Comparison of accumulative cash flow for pit A and pit B

prior design other than past experience in similar rock. It is necessary, therefore, to expect major slope failures from these slopes. Numerous authors have discussed the importance of these failures (11) (12) (13) (14) (15). Technologically speaking, there are a large number of different types of slope failures; but as far as pit operations are concerned, there are only two types — catastrophic failure and progressive failure. The first type, catastrophic failure, is the more dangerous to open pit operations. The slide gives little warning of its existence and then fails at high speed. The progressive failure, on the other hand, gives plenty of warning and fails by gradual decrepitation of the slope face and the pit crest. This type of failure can possibly become catastrophic in its later stages of development. In all cases of major slope failures, some form of warning is given by the rock. Today, it is essential that an open pit operation have someone on the operating or engineering staff who can spend some time each week patrolling the mine looking for signs of such slope instability.

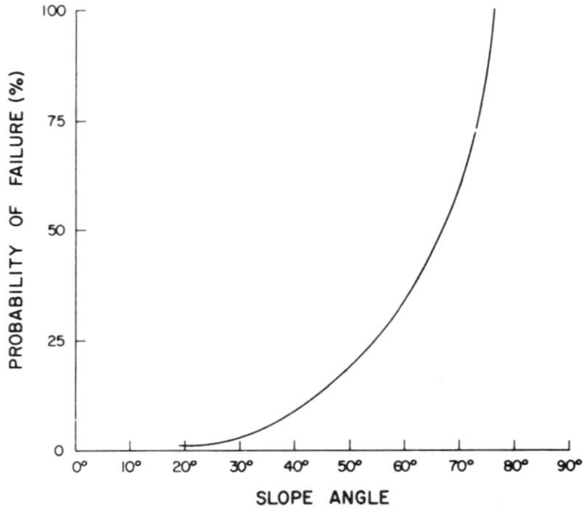

Fig. 5 —Example of failure probability chart for use in risk analysis

Ignoring a Failure

When signs of instability in a slope are noticed, what steps should be taken? Of course the simplest and most dangerous thing to do is to ignore it. In these days of modern technology and ever increasing sophistication in the mining industry, such action is inexcusable. If a failure ensued, and an investigation followed, the reaction of the investigating committee could be similar to that of the tribunal set up to inquire into the Aberfan disaster in Wales, which strongly and unanimously stated that the disaster could have been avoided if those involved had not "been oblivious of what lay before their eyes . . .". The indictment goes on to cite that "the Report tells not of wickedness but of ignorance, ineptitude and a failure in communications . . . and a failure on the part of those having knowledge of the factors which affect dump stability to communicate that knowledge and to see that it was applied". From a legal standpoint, there are three main areas where the open pit mine operator has to consider the pit slopes and be cognizant of the possible damage they can cause. First, there are the legal statutes regarding the slopes and their possible effect on adjoining land owned by another party. These statutes vary from country to country, and from state to state within each country. Secondly, contractual liability, where a contract may have been written upon the acquisition of the property which may contain an agreement to hold a certain slope angle or to provide lateral support to the neighboring ground to prevent damage. Thirdly, there is tort liability, which is the contractual right that society gives to the individual to protect him from negligence or harm from a third party, in this case the mining company operating the open pit.

In summary, to ignore a major slope failure in its early stages is dangerous, and could result in a series of lawsuits as well as unfavorable economic consequences to the operation itself.

Removal of a Failure Zone

The second course of action that could be taken if a major slope failure develops is to remove the slide in toto. This course of action must be examined very carefully both from operating and economic standpoints before it is undertaken. First, it is imperative that enough background data be available on the slide to enable the engineering and operating staffs to satisfy themselves, and management, that there is enough time to remove the slide. In order to do this, some long-range predictions of the behaviour of the slide have to be made. Admittedly, this is extremely difficult to do, but provided that a monitoring system has been used, it should be possible to come up with an approximate earliest date of failure prediction. Confidence limits will have to be applied to this to assist management in its decision. If it is thought that only a small amount of time is left before failure, and if mining is authorized with the attitude that to remove some material is better than not to remove any at all, it must be realized that the chances are that the failure date is only being delayed by a matter of days, weeks, or months. Also, when a failure zone is mined out, this is **not** the same as unloading, which will be dealt with below. In other words, in removing a failure zone the mining sequence is such that the areas of the slide that most affect the failure are not necessarily removed first.

The following considerations have to be made by the operating staff: Does the area have to be stripped anyway? If it does, can the operation afford to put equipment to work in the area with a possible consequent drop in production and a slowdown of stripping in possibly more important areas? How does this extra stripping affect the cash flow of the project? Does it increase the deferred stripping fund? If the area does not have to be stripped in the original overall plans for the pit, it may be better to find some way to live with the slide, as will be discussed below. If the decision is made to mine the slide out, the following points must be examined: First, the new slope formed must be designed to be stable and: secondly, the effect of removing this wedge of material from the slope on the rest of the pit perimeter must be examined. If a tangential stress exists in the pit walls, there will almost certainly be some migration of the areas on either side of the new slope in toward the newly mined area.

Controlling or Unloading a Failure Zone

The third course of action is to control or unload the slide. A good monitoring system is essential prior to the decision to carry out control measures, during the operation, and after the control measures have been put into effect. Good background information on the behaviour of the failing mass, its size, shape, and structural control are essential. A careful examination of the possible mechanics involved in the failure should be made. A feeling for the directions and magnitudes of the movement is essential to determine which areas of the slide should be removed, and what tonnages are involved. As a general rule, this approach can only be used where the stripping necessary to remove the slide is not included in the original mine plans, and where important installations have to be protected. Once again, as with removal of a failure area, careful consideration must be given to the effect on mine plans and the economic cash flow. Can equipment be spared for the job and, if so, can it be done in time? The engineers concerned must be fairly certain that their efforts will stabilize the failure zone. Support measures can be considered, such as buttressing with waste material at the toe if some operating room can be spared, or the use of long, pretensioned cables in hard rock. If an attempt is to be made to control or unload a failing zone, the pit operators must be certain that the investment is worthwhile and that the required result is likely to be achieved.

Living with a Failure

The fourth course of action is to live with the slide. Many open pits have been faced with the situation where a failure starts in a zone that is considered to be the final pit limit so that the economics of carrying out any mining in the area are not really justified. In this case, a careful analysis of the outcome of a possible failure must be made. If necessary, it may be advantageous to drive alternate haulage ramps in case the existing ramps are cut off. A very close monitoring program is recommended, and an attempt should be made to predict the date of failure. Of course, the chance of a major slope failure occurring in a part of the open pit that does not matter is fairly small. Normally, a failure is likely to cover up ore reserves or threaten an active mining area. As has already been discussed above, attempts can be made to remove the failure area, or some attempt can be made to control the moving mass. However, if time is not available to remove the failure, and control measures are only likely to be partially successful, some attempt can be made to mine the ore from the toe area before the failure occurs. Most techniques of living with a slide are unique to the situation and the mine concerned. In one case, a fairly large open pit in the U.S.A. was faced with the problem of very incompetent, highly fractured, and highly altered rock overlain by an alluvium overburden. A number of slope failures had occurred throughout the life of the mine, but the operating personnel found that by unloading a given tonnage from the top of the slide area, that they could then go in at the toe and remove approximately the same tonnage of ore. The slide was maintained in balance by this method. At another open pit, a progressive failure started in the rock overburden and threatened to cover approximately 3 million tons of fairly high grade ore in the toe. The slide was very slow moving and relief was being shown by the breakup of the slope face. A plan was devised in which a saturation of monitoring points was installed on the slide and then mining equipment was moved in at the toe to remove a slice of ore. The slide naturally accelerated during the mining. Once the mining of the slice was completed the equipment was removed and the slide was allowed to settle down. Once the velocity had returned to a reasonable rate, the equipment was moved in again and another slice taken. The work at this mine is far from complete, but what has been done to date has shown, that, in a situation where the cost of stripping the waste material and the slide area would have been excessive, that it is possible to recover ore which otherwise would not have been mined for economic reasons.

The Improvement of the Stability of Slopes

The three most likely areas that can offer some improvement in the stability of slopes are 1) the recognition of groundwater problems and the effective treatment of them, 2) the recognition of the damage caused to final slopes by blasting, and the careful design of blasting techniques to reduce this damage, and 3) the possibility of the use of artificial stabilization techniques, especially in critical areas.

Hydrology

Groundwater is one of the greatest contributors to slope instability problems. A prime example of a major slide that was virtually caused by a change in the groundwater condition in the rock mass was the failure at the Vaiont Dam in Northern Italy. The Vaiont Dam was a double curvature arch dam, some 292 m. high, located on a tributary of the Piave River, north and west of Venice. When completed in 1960, it was the highest arch dam in the world. In 1960, a sizable landslide involving some one million tons of rock occurred in the left dam abutment, a short distance above the dam itself. Following this, the reservoir was shut down for construction of a bypass tunnel through the right valley wall to ensure connection with the upper end of the reservoir in the event of further slides. In October 1963, a giant landslide occurred on the left valley wall over a width of some 2.4 km. Over 405 million tons of rock slid into the reservoir, displacing the water and forming a huge wave that was 230 m. above the top of the dam when it hit the opposite, or right wall of the valley. The wave then passed some 200 m. over the top of the dam, followed down the winding canyon, and spilled out into the valley of the Piave where it engulfed the village of Longarone causing the loss of 2,050 lives. Remarkably, the dam stood. The slide lasted approximately two to three minutes. A simplified description of the cause of the slide is as follows:

As the water in the reservoir rose, it caused an increase in the pore water pressure in a contact plane separating a limestone formation from an alternating shale and limestone series overlying it. This increase in pore water pressure reduced the effective shear strength of the plane causing the upper formation to slide along the plane and into the reservoir. This is a very dramatic example of the influence of water on a rock mass. Many natural slopes have a very low safety factor, so that minor changes in the site conditions may cause instability. In open pit mining we upset the natural balance by excavating a slope and consequently changing the hydrological conditions in that slope.

A careful hydrology study, during the exploration phase for a new mining property, is a very worthwhile investment since it will delineate any possible future water problems in the slopes and allow the necessary drainage and instrumentation systems to be designed. Drainage tunnels, pumping well systems, and horizontal drains are some of the methods that can be used to remove the water from a slope. In certain cases, stabilization of failing open pit walls has been achieved solely by the installation of horizontal drainage pipes. The installation of such horizontal drainage pipes is very inexpensive; however, drainage galleries and pumping systems can become quite costly and require careful consideration before installation. The control of surface runoff, by carefully designed drainage ditches and sumps, is also essential.

Blasting

One of the major problems that arises from the excavation of rock slopes is that the very material we are trying to keep stable is seriously damaged when it is subjected to the very high seismic shocks caused by blasting. Most open pit operators are familiar with backbreak from blasts, but most people only consider the visible breakage behind the last row of holes of the blast. But how far back does the damage really extend, when damage is defined as actual deterioration of the rock structure itself? In one recent example, a tunnel was driven in the supposedly highly competent rock of an open pit wall. Spiling had to be used to drive the tunnel in the 80 m. closest to the pit wall. It was felt that at least 50% of this damaged rock was caused by the seismic shock from the large blasts in the pit. The slopes in an open pit that need the most protection from the effects of blasting are the final walls since it is here that steepening is most likely to be carried out and, hence, the best rock charactertistics are required.

How can blasting be improved to cut down wall damage? The three most likely methods are 1) to reduce the amount of charge per delay to a minimum, 2) to use presplitting on the final walls, and 3) to use a buffer zone between the presplit and the main bench shots. By reducing the charge per delay, the amplitude of the seismic shock is kept small. The delay system used should also be examined to ensure that the various charges do not amplify each other and cause a high amplitude harmonic that in itself can cause damage. The optimum loading per delay is best achieved by experimentation, since it will almost certainly be a unique value for any given situation.

Presplitting is possibly one of the best tools presently available to the open pit operator to improve the condition of his final walls. Presplitting is basically the creation of a fracture between the blast and the final wall by shooting a series of decoupled charges in closely spaced holes. Experimentation has to be carried out to determine the optimum hole size, spacing, and hole depth, but typical examples are 3-in. dia. holes spaced on 3-ft. centers. Good results have also been obtained in some open pits with 6-in. dia. holes on 7-ft. centers. The decoupled charges can consist of special presplitting powder, or spaced sticks of dynamite strung on Primacord, and in at least one case, very successful presplitting has been done with one-pound boosters strung at 30-in. centers on Primacord in 6-in. dia. holes on 6-ft. centers. One of the problems with presplitting on high benches with widely spaced holes is the inability to pull a good clean toe. Experiments are currently underway to examine the effect of presplitting double benches, mining on single benches, and ending up with a safety and access bench every two benches.

A buffer zone can be introduced between the main blast and the presplit. This consists of a series of holes more closely spaced than the main blast holes but on larger centers than the presplit itself, and blasted sequentially between the presplit and the main shot. The zone creates a shock absorbent area, a buffer between the main blast and the final wall cut by the presplit; hence, further reducing the transmission of shock through the presplit to the wall.

On a roadcut for the U. S. Interstate Highway System, slopes in limestone of up to 500 ft. high and at angles of up to 74° have been successfully cut using presplit holes as long as 200 ft. on 3-ft. centers in conjunction with a buffer zone.

What does presplitting cost? A recent study on an open pit involving the removal of some 155 million tons of material, both ore and waste, showed that if presplitting were carried out on the final walls, which averaged 800 ft in height, that the overall mining costs for the entire 155 million tons would be increased by between 1.1¢ to 1.5¢ per ton of material removed. This increase includes both capital and operating costs.

Returning to the hypothetical pit shown in Fig. 1, it is a very simple matter to calculate the value of using presplitting on the final wall. If it is assumed that mining had been carried out with 45° slopes, and that presplitting techniques can only increase the final slope by 1° to 46°, is the investment worthwhile? The total tonnage, ore and waste, to be removed from the pit will be 201,553,875 tons at a mining cost of $0.265 per ton instead of $0.250 per ton, assuming that the presplitting will increase the mining cost by 1.5¢ per ton. On the other hand, the one-degree increase in the slope angle will make an additional 803,570 tons of ore available at an $8.50 credit per ton to the mine. The increase in total mining cost will amount to $3,023,308, and the increased income from the ore will be $6,830,345, resulting in the mine being $3,807,037 better off if presplitting is adopted to increase the final slope angle by only one degree.

Artificial Stabilization

The third area of possible stability improvement is the use of artificial stabilization techniques on slopes above critical installations, such as crushers or on slopes on which critical installations are located. Some of the artificial stabilization techniques that can be used are:

1. Toe Buttresses or Retaining Walls. These can consist of rock backfill or of a massive reinforced concrete buttress placed at the toe of the failing zone to prevent or limit movement of the toe. These forms of stabilization have been very successful in small soil and rock slides, but to date they have not been used in controlling large rock slides.

2. Bolting. Long rock bolts, or pretensioned cables, have been used in a large number of stabilization projects with considerable success, although care must be exercised in locating such installations if they are to be successful. This can be an extremely expensive way of stabilizing a slope. A detailed engineering and economic study of the situation is recommended to determine if it would not be less expensive to move the installation and allow failure to occur.

3. Grouting. Chemical or cement grouting are also possible stabilization techniques. Although once again, they must be applied with extreme caution. If a large rock mass is involved, these methods can become very expensive. A second factor to be considered is that the injection, of what is initially a fluid, into the joints and fractures could in itself precipitate failure. A detailed engineering and economic study is also recommended in the use of either of these techniques.

4. Others. Various other methods have been suggested from time to time, although few, if any, have been tried in the field. These include freezing the mass by the same techniques used for shaft sinking in saturated materials; electrochemical methods in which the electropotentials in the ground are artifically altered in order to change or increase the ionic bonds between the various minerals in the rock, hence, supposedly increasing the overall stability.

Most stabilization projects are expensive, and in at least one such project in an open pit, five times as much money was spent on stabilization as it would have cost to move the installation that was being protected.

Waste Dumps, Leach Dumps, and Tailings Dams

Once the ore leaves the open pit, the open pit miner sometimes tends to think that what happens to the ore and the inevitable tailings is the metallurgist's problem. However, he cannot avoid some of the responsibility for the siting of the tailings dams and the actual emplacement of the tailings. A number of accidents in past years involving tailings dams have been reported in the international press, and with the ever-increasing public interest in ecology and pollution control, a considerable increase in the careful siting and design of tailings disposal systems is inevitable. It is unfortunate that considerable sums of money have to be spent on what is essentially a waste product, but these costs are minimal compared to the damages that would have to be paid out and the ill feeling generated if a repetition of what happened at El Cobre in Chile or at Luanshya in Zambia occurred in, or near a populated area. At El Cobre a number of tailings dams were liquefacted by an earthquake, and the resulting mud-flow engulfed a village causing great loss of life and an immense amount of property damage. In the case of Luanshya, a breach occurred in a tailings dam and a large proportion of the contents of the dam were deposited in a nearby river. If such an accident occurred in a populated area or involved a river that was a water supply, it could cause rather extensive repercussions.

With the current status of technology, the design of tailings dams is a far more exact engineering exercise than the design of rock slopes in an open pit. The dam is a man-made structure over which he has considerable control and which is not plagued by the vagaries of jointing, fracturing, rock types, etc., that occur in a rock slope. Tailings dam design is fairly inexpensive, and many reputable organizations are well qualified to carry out the necessary work.

The siting of waste and leach dumps is of extreme importance to the open pit operator. First, he must ensure that the dumps are not sited in such a position that they will have to be removed in later years when the ability to mine lower-grade ores at a profit will necessitate mining near, or under these dumps. Secondly, the dumps must be placed so that they do not cause a surcharge load on the pit slopes, and in the case of leach dumps, a groundwater recharge source.

Open Pit Slope Stability Studies

There are three methods of approaching slope stability engineering in open pits; namely, 1) a total in-house group, 2) the exclusive use of consultants, or 3) a combination of both, involving a certain amount of in-house capability and supplementing it by the use of consultants for specific problems. The first approach is both difficult to achieve and expensive. Good, well-trained, slope stability engineers with practical experience are difficult to find; and a large backup team consisting of geologists, mining engineers, hydrologists, etc., are also necessary. The second approach is also expensive and suffers from the serious problem that a slope stability study of an open pit is a continuing, long-term project and that very few, if any, consulting companies will sign long-term contracts for such work. The third method is the best since it involves both considerable in-house capability utilizing men familiar with the open pits concerned, the methods of operation, and the company policies involved and consultants for specific problems where the in-house team does not have the necessary capability to handle the particular problem.

The costs of slope stability work vary tremendously, and no rigid guideline figures are available. However, there is little doubt that they pay off. For example, the slope stability work leading up to the prediction of the 1969 slide at Chuquicamata Mine cost about 1% of what would have been lost if the slide had been ignored and had halted production for any length of time (16).

The institution of a slope stability study at an existing open pit does not necessarily mean that the group will have to start from scratch. A large amount of the basic data necessary for slope design is already available in geological and engineering records. Data will be available, although not necessarily in the form, or in the detail required by the slope stability engineer, on geological structure, rock types, hydrological conditions, old failures, etc. The slope stability engineer can improve his lot considerably by "educating" the geological, engineering, and operating departments to collect data for him. For example, during geological exploration drilling, much useful information for slope design can be collected from the drill holes and the cores. This data collection does not take long, and can be done most easily by the exploration geologist. In cooperation with the engineering and operating department, the slope stability engineer should be kept posted on the occurrence of minor bench failures. These can then be back-analyzed to provide design data, in the form of joint shear strengths, etc. from what is essentially a full-scale field test. Once management and the operating staff accept slope stability as a necessary part of the operation, it will be possible to design the open pit slopes from an engineering basis to optimize the profitability and operational safety of the mine.

CONCLUSIONS

Slope stability engineering is no longer a theoretician's paradise of little value to the industry, but is a practical design tool that can be used in the daily planning and operation of an open pit, often with enormous economic benefit. We are now in a position to better understand the mechanics of rock slopes and, hence, can deal with rock as an engineering material. The mining industry must now accept that the old rule of thumb of 45° slopes for an open pit is an outdated concept that can adversely affect the economic future of a mine. The technology of slope stability will play an important part in the future mining of low-grade, previously marginal orebodies, which will have to be exploited if the mineral needs of the world are to be met.

REFERENCES

1. Black, R.A.L., and Jennings, J.E., "Factors Affecting the Angle of Slope in Open-Cast Mines," Preprint No. 634011 of SME of AIME, February, 1963.

2. Rausch, D.O. and Soderberg, A., "Pit Planning and Layout," Surface Mining, SME of AIME, 1969, pp. 141-165.

3. Hoek, E. and Pentz, D.L., "Review of the Role of Rock Mechanics in the Design of Opencast Mines," 9th Commonwealth Mining and Metallurgical Congress, London, October 1969.

4. "Lessons of Aberfan," Mining Journal, Vol. 169, No. 6886, London, August 11, 1967.

5. Kinnear, J.C., "Rock Mechanics and Slope Stability in Management Terms," SME of AIME, Washington D.C., February, 1969.

6. Gill, D.K., "Some Economic Aspects of Open Pit Mine Planning," Arizona Section Regional Meeting, AIME, Tucson, Arizona, December, 1967.

7. Erickson, J.D., "Long Range Open Pit Planning," Mining Engineering, April, 1968.

8. Merrill, R.H. and Rausch, D.O., "Slope Stability," Surface Mining, SME of AIME, 1969, pp. 773-790.

9. Brawner, C.O., "Stability Investigations of Rock Slopes in Canadian Mining Projects," 2nd International Conference on Rock Mechanics, Belgrade, 1970.

10. Call, R.D., "Design of Open Pit Slopes," paper presented at SME of AIME, Las Vegas, September, 1967.

11. Broadbent, C.D., "Slope Stability Program of Kennecott Copper Corporation," Transactions of Society of Mining Engineers, June, 1969.

12. Brawner, C.O., "Slope Stability in Open Pit Mines," Western Miner, October, 1966.

13. Black, R.A.L. and Hoek, E., "Status of Rock Mechanics as Applied to Mining," Ninth Symposium on Rock Mechanics, Colorado School of Mines, Golden, Colorado, April 17 to 19, 1967.

14. Plewman, R.P., "The Basic Economics of Open Pit Mining," Symposium on the Theoretical Background to the Planning of Open Pit Mines with Special Reference to Slope Stability, Johannesburg, September, 1970.

15. Halls, J.L., "The Basic Economics of Open Pit Mining," Symposium on the Theoretical Background to the Planning of Open Pit Mines with Special Reference to Slope Stability, Johannesburg, September, 1970.

16. Kennedy, B.A., et al, "A Case Study of Slope Stability at the Chuquicamata Mine," Chile," Preprint No. 70-AM-81, SME of AIME, February, 1970.

Chapter 3

GEOLOGIC FACTORS CONTROLLING SLOPE STABILITY IN OPEN PIT MINES

by

F. D. Patton

Associate Professor of Geology, University of Illinois

and

D. U. Deere

Professor of Civil Engineering and Geology, University of Illinois

INTRODUCTION

A close relationship exists between the geologic investigation and the stability analysis of the slopes of an open pit mine. One significant result of this relationship is that the stability analysis is no better than the assessment of pertinent geologic factors that enter into the analysis.

In some slope stability problems the engineering solution becomes obvious once the geologic framework is established. In other cases, establishing the geologic framework is only the first stage in the process of analysis. However, even in these cases the geology should not be neglected once the analysis is made. The geology should be periodically reviewed when new slopes are excavated. This will help ensure that the basic assumptions for the analysis have not changed and will permit changes when new conditions are encountered.

In recent years improved methods of stability analysis have been accompanied by advances in laboratory testing techniques and new knowledge of the shear strength of rock discontinuities. At the same time there have been improvements in field geological descriptions.

As more experience is gained in analyzing and predicting the stability of rock slopes, the importance of certain geologic factors has become more apparent. Some of these factors were described in a paper given at the Open Pit Mining Symposium in Johannesburg, South Africa in September 1970 (1). This paper includes many of the points covered at that meeting but with additional emphasis on the significance of the major geologic structures. Other papers given at the South African symposium also considered geological and hydrogeological factors in slope stability analysis (2) (3) (4) (5) (6).

This paper first describes the approach to a slope stability problem including assessment of the relative roles of field geology studies, laboratory testing, stability analyses, and continued slope monitoring studies. Next, is described a framework which has been found useful for classifying or separating the several types of slope stability problems encountered in open pit mines. The principal geologic factors in slope stability problems in rock are discussed and brief comments are made on how these factors distinguish stability problems in rock slopes from those normally encountered in soil slopes. The over-riding control of major geologic structures, especially faults, in most slope stability problems is emphasized. In this regard the relative importance of statistical studies is contrasted with knowledge of the positions of key structural features shown on maps and the projected positions of these structures in cross sections of future mine slopes. Finally, several elements in the stability of rock slopes which deserve special consideration will be described.

APPROACH TO SLOPE STABILITY PROBLEMS

The purpose of soil and rock mechanics studies of open pit mines is to ensure reasonable stability of the slopes in the most economical way. These studies generally include (a) detailed geological mapping, (b) determination of the groundwater conditions in the open pit area, (c) determination of the mechanical properties of the soil and rock by both laboratory and in-situ field testing, (d) analytical studies of the slope stability for given slope angles and in different portions of the mine, and (e) monitoring the behavior of the slope with continuing geologic surveillance and appropriate instrumentation.

Depending on the results of the above studies, it may be necessary for safety and stability to design flatter slopes or to provide certain remedial measures. These could include wider benches at particular locations, drainage wells or drainage galleries, tie-back anchors or deep rock bolts, and surface treatment of slopes to prevent loosening, ravelling, or erosion.

Conflicting requirements are always present in stability studies of open pit mines. From the viewpoint of economy, steep slopes and minimal remedial measures are desired. However, from the viewpoint of safety and stability (and therefore indirectly, economy), the slopes should be as flat as possible and numerous remedial measures may be required or desirable. It is difficult for management to select the best excavation procedure until the stability of various possible slopes has been evaluated.

The primary responsibility for those conducting soil and rock mechanics investigations for stability studies of open pit mines is to prevent major slope failures which could result in loss of life and/or serious production stoppages. In order to meet this responsibility a rather detailed initial study is required followed by continued mapping and surveillance of the open pit slopes and exposed geological features and careful monitoring of the groundwater conditions.

Analysis of the test results, groundwater levels, and geological conditions must be kept up to date so that possible dangerous slope failures can be anticipated and prevented by applying appropriate remedial work where necessary. It becomes the duty of the geologists and soil and rock mechanics experts to anticipate potentially dangerous areas long before they start to fail and become obvious to the operating personnel. Many of the investigated areas will be shown by the studies to be relatively safe; others, however, may be shown to be so critical that immediate action is required.

It is best if the geologist has a working knowledge of the shear strength of natural materials and of methods of stability analysis. In this way it is more likely that the truly significant geologic factors will be recognized and mapped. In a similar vein, Müller and Hofmann (2) recommended model tests at an early stage in the investigation to permit better selection and interpretation of the geologic field data. In any event, close cooperation between the field geologist and the personnel making the analysis is essential.

The role of the geologist is critical in both the collection and presentation of data for it is easy to collect more data than can be used in an analysis. This can result in the analyst misdirecting his emphasis. Hence, only the most pertinent data should be presented. Yet any limitation on the quantity of data presented to the analyst will reflect the prejudices of the collector. As a result, experienced geologists are essential in this phase of a study, for the geologist's prejudices must be in agreement with and correctly anticipate the actual mode of failure of the slope (should a failure occur), the location of the critical materials, and the actual distribution of fluid pressure within the slope. It is therefore the job of the geologist to examine a large number of geological parameters and reduce them to a few significant ones. In many cases the final determination of which parameters are significant must await the outcome of the preliminary stability analysis. At this point new field investigations may be required.

Müller and Hofmann (2) have noted that in the "assessment of the stability of a slope the interpretation of the geological data is not only one of the most important but unfortunately one of the more unreliable and indeterminant steps." They also noted, and it has been our experience, that stability calculations using strength and fluid pressure parameters based upon geological data are often made to several significant figures. Occasionally these calculations may include consideration of errors resulting from different magnitudes of the selected parameters. Yet rarely is thought given to the major error which may result from an unrealistic assessment of the relative significance of the parameters or from a wrong choice of parameters. Knowledge of regional geology, the origin of the local materials, and the stability and failure of natural slopes often helps to prevent such major errors in the assessment of slope stability.

In conclusion, we must try to keep in mind and live with the fact that nature is often more complex than it first appears. The state-of-the-art of rock mechanics and our ability to locate every geologic discontinuity and to analyze their effects are not such that absolute security can be guaranteed. This is particularly true for the complex geology found in the vicinity of many ore bodies. Slope failures may still occur. However, careful studies should greatly diminish the number and seriousness of slope failures and allow the open pit mine to be developed as safely and economically as possible.

THREE PRINCIPAL TYPES OF SLOPE STABILITY PROBLEMS

Many different slope stability problems can be encountered in the excavation of an open pit mine. These problems have different origins and commonly have appreciably different effects on the design and operation of the pit. We have found it convenient to group the typical mine slope stability problems into three categories:

Type 1 — local slope failures involving a single bench,

Type 2 — large-scale wedge failures involving several benches,

Type 3 — failures in sheared and decomposed rock which may involve several benches.

These types of slope stability problems are illustrated in Fig. 1 and are described below.

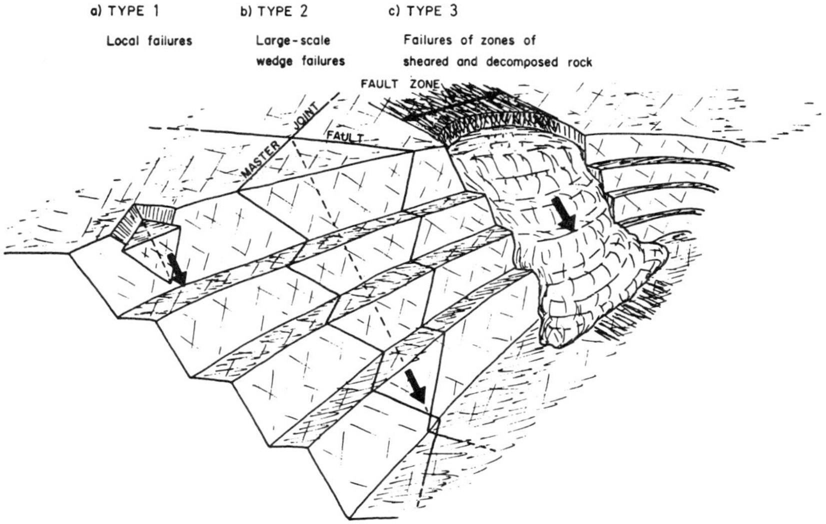

Fig. 1 — Three principal types of slope failures in open pit mines.

Local Slope Failures (Type 1)

Local slope failures of small rock masses along one or more joint planes, such as the one illustrated in Fig. 1a, are common in any large excavation. They are almost impossible to eliminate completely without the use of excessively flat slopes or a costly slope-support system such as gunite applied over wire mesh that is secured to rock anchors. In most cases, if a few local failures did not occur, one might conclude that a slope is overdesigned and is probably too flat.

These local failures are considered to extend a vertical distance of less than the height of one bench. Such failures would not usually appreciably influence the overall mining operations for more than a few days at the most. The worst problem presented by these local failures is that of the hazard to men and equipment. However, in some present-day mines the cost of a single piece of equipment may exceed 20 million dollars and in this case it would be possible for a local slope failure to appreciably change the entire economics of the mine operation. Local failures may also be significant adjacent to conveyor belts, tunnel portals, or other relatively permanent mine structures.

In general, good slope design will minimize local failures but it is unlikely that many mining operations could afford to eliminate all such failures. Safety considerations may require close observation and monitoring of certain individual bench slopes by field personnel during the following excavation. It is not essential that one have extensive training in field geologic methods to recognize the factors leading to these local failures. Experienced mine personnel often recognize the geologic conditions leading to these failures at an early stage and adjust their mining operations accordingly.

Large-Scale Wedge Failures (Type 2)

The presence of two or more through-going discontinuities, such as a prominent bedding-plane joint or master joint combined with a fault, may lead to the situation illustrated in Fig. 1b. This geologic condition is potentially much more hazardous than the local failure. The hazard results because: (a) a much larger rock mass may become involved in the slide; and (b) the geologic conditions leading to the failure are often difficult to detect in advance, hence, there is less chance that the failure will be predicted. The difficulty in detection and prediction occurs where two adverse geologic structures are separated by hundreds of feet at the surface and where the influence of rock weathering or other overburden may make them hard to detect. Careful geologic field work may be required to find these significant structures; the structures must then be projected to anticipate their influence on future mine slopes.

Conditions conducive to a large-scale wedge failure can cause mine operators and workers to develop a false sense of security as there may be little or no warning of an imminent slope failure until the line of intersection of the two geologic structures is exposed or "day-lighted". In the examples shown in Fig. 1b, the failures probably would not occur until the excavation had reached to within a few feet of the intersection of the fault and the master joint. A failure of such a rock wedge could stop mining operations for months and could conceivably change the economics of the entire operation. For these reasons it becomes one of the prime obligations of the project geologist to locate the principal through-going structures and to anticipate such failures before they occur.

Failures of Zones of Sheared and Decomposed Rock (Type 3)

Where the mine slopes encounter wide fault zones containing sheared and decomposed rock, the slope design used for the sound rock will have to be modified to account for the weaker material.

The geologic evidence of such weak materials is usually easier to see than that for Type 2 failures. However, like Type 2, the slope may not fail until the excavation has proceeded considerably below the elevation where the weaker materials were first encountered.

Failures of Types 1 and 2 are best analyzed by the techniques developed in rock mechanics, whereas failures of Type 3, as illustrated by Fig. 1c, can usually be treated by the more conventional methods developed in the field of soil mechanics. Adverse groundwater conditions are commonly encountered where the geologic structures are such that Type 2 and Type 3 failures may develop.

Summary of the Principal Types of Stability Problems

Conditions leading to failures of Types 2 and 3 are important enough to be major considerations in the layout of an open pit mine. Thus, the main objective of an exploration programme would be to identify and describe the major fault zones, areas of weathered rock, and all through-going joints, faults, and beds of weak rocks. The pit layout and slopes based upon the major geologic structures would then be modified to whatever extent is possible by the characteristics of the secondary discontinuities in order to minimize the occurrence of the Type 1 failures.

Since a mine pit has several sides, the through-going geologic structures are likely to intersect more than one side. Therefore, the most favourable position and slope for one side of the pit may result in stability problems for another side. In these cases the slope stability problem becomes one of optimization of the pit layout geometry.

DIFFERENCES IN STABILITY PROBLEMS FOR SOIL AND ROCK

Two main groups of geologic factors distinguish slope stability problems in soil from those of rock. One group of geologic factors is related to the omnipresent structural defects found in rock masses and the special strength problems that result, whereas the other group is related to special groundwater conditions which are more commonly associated with rock masses than with soil. The critical groundwater conditions are often a direct consequence of the presence of structural defects. In general, rock masses are best considered as possessing anisotropic strength, permeability, and deformability characteristics to a much greater and more significant degree than do soils.

SIGNIFICANT GEOLOGIC STRUCTURES

Oriented Planes and Zones of Weakness

The distinguishing feature of slope stability problems in rock is that the failure planes conform so closely to pre-existing planes of weakness. In the average slope stability problem in soil, but admittedly not in all cases, one assumes that the soil has a relatively equal strength in all directions. However, in rock the strength along a discontinuity may be only a small fraction of the strength of the intact material. Hence, the need to locate and establish the orientation and strength properties of the critical discontinuities in rock is obvious. There is seldom an equivalent need to establish the location of a single plane in soil slopes. Residual soils would, however, be important exceptions to this generalization.

In addition to changes in strength, discontinuities and weathering can result in equally large changes in the modulus of deformation and the permeability. These changes are all possible within the same rock type resulting in larger variations than are commonly found within a single soil type.

Variations in Strength Due to Irregularities Along Rock Discontinuities

Seldom does one have to consider the nature of the irregularities along failure surfaces in soils. Yet in rock the irregularities along a fault or joint surface can mean the difference between stability and failure of a mine slope. The presence of irregular rock surfaces (such as those shown in Fig. 2, 5 and 6) can result in the possibility of different modes of failure along the same rock surface. Depending upon the normal stress level, the irregularities can be overridden or sheared off. Thus, different shear strength-normal stress relationships exist at different stress levels for the same rock surface. These and other effects have been described by Patton (7) (8) and others.

Large Losses of Shear Strength Due to Displacements

The intact shear strength of rock is much larger than the "undisturbed" strength of soil having the same mineralogic composition. However, the residual shear strength or the strength obtained after large displacements is roughly the same for soil as for a relatively flat joint surface in the equivalent rock. Hence, the loss of strength with displacements can be several orders of magnitude greater for rock than for soil. This relationship is shown in Fig. 2.

Fig. 2a shows shear strength vs. displacement curves for a rock and a soil with the same mineralogy which were tested under the same normal stress σ_{n_1}. It is obvious that the peak strength for the rock will be many times that for the soil, yet as displacements continue the residual strength of both materials is approximately the same. Fig. 2b is a summary plot that one could prepare from a series of shear strength tests on a series of identical specimens of rock and a series of identical specimens of soil, both series having the same mineral composition.

The maximum and residual shear strengths are plotted for each test at the appropriate level of normal stress. The results for the two tests from Fig. 2a are shown plotted along the vertical line for σ_{n_1}. Fig. 2a and 2b are two ways of illustrating that the loss of strength possible in rocks is of a much greater magnitude than that for most soils. The large loss of strength with small displacements that is characteristic of many rock discontinuities is one reason why observations of small displacements are so significant in slope stability studies. It is also why failures of rock slopes often give less advance warning than is common for failures in most soil slopes.

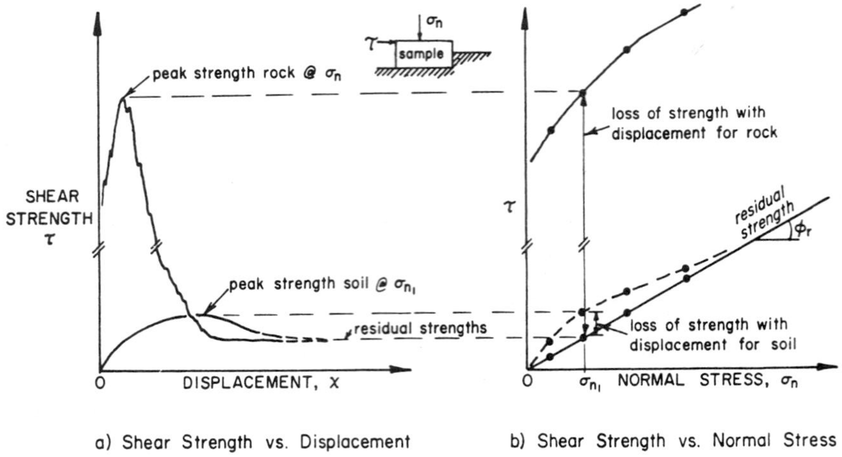

Fig. 2 — Comparison of the loss of strength with displacements between soil and rock.

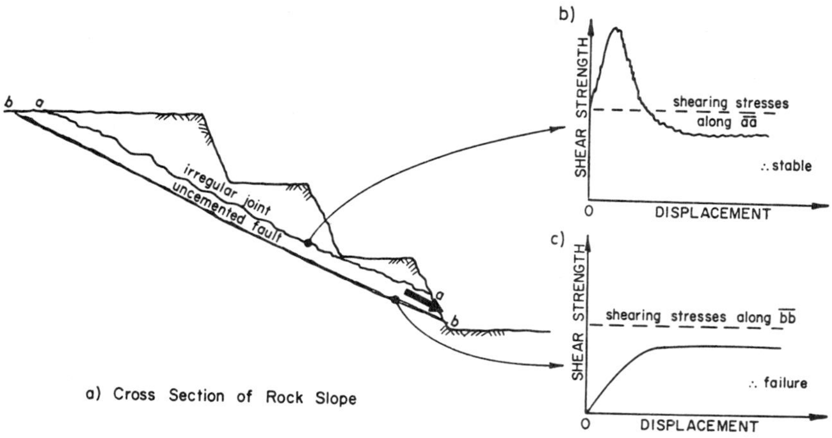

Fig. 3 — Significance of pre-existing displacements along faults.

The low residual strengths obtained along rock surfaces which have undergone considerable displacement is one of the reasons why uncemented faults and shears are so significant in slope stability problems. Fig. 3a is a cross section of a rock slope with an irregular joint and an uncemented fault, both with unfavourable orientations. In this example porewater pressures are assumed to be negligible. Fig. 3b shows the shear strength diagram for the irregular joint. Since the shearing strength at small displacements is shown to exceed the shearing stresses, the joint will remain stable. However, Fig. 3c shows that when the pit is excavated deeper to expose the uncemented fault the shearing strength at all displacements will not be sufficient to resist the shearing stresses and a failure of the slope will result.

The geologist must search for faults or shear zones having low shear strengths due to previous displacements. The search is aided by the knowledge that faults or shear zones are characteristically associated with particular geologic environments. These include:

(a) faults subparallel to or in secondary or conjugate alignment to regional faults,

(b) bedding plane faults in shales where they are interbedded with other rock types,

(c) foliation shears in schists and slates, and

(d) foliation shears along micaceous bands within gneisses.

The faults or shears described in items (b) and (c) above are particularly common in folded or inclined sediments and adjacent to thick layers of a relatively less deformable rock such as sandstone, basalt, quartzite, or amphibolite.

Examples of faults or shear zones in several of the geologic situations described above are shown on Fig. 4.

Irregularities Associated with Faults and Shear Zones

Different types of irregular surfaces are commonly associated with different sets of joints and faults. In addition, the nature and effect of the irregularities may be different in different directions on the same rock surface due to an oriented structure developed on the surface.

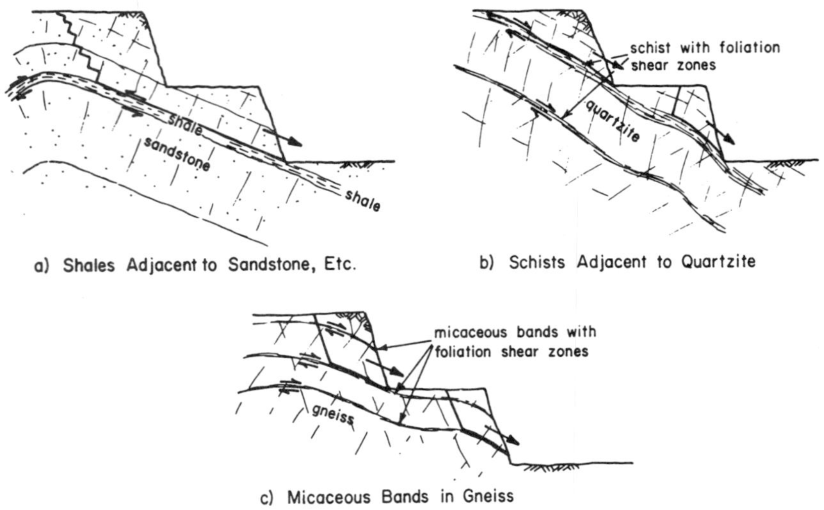

Fig. 4 — Typical occurrence of shear zones in layered rock.

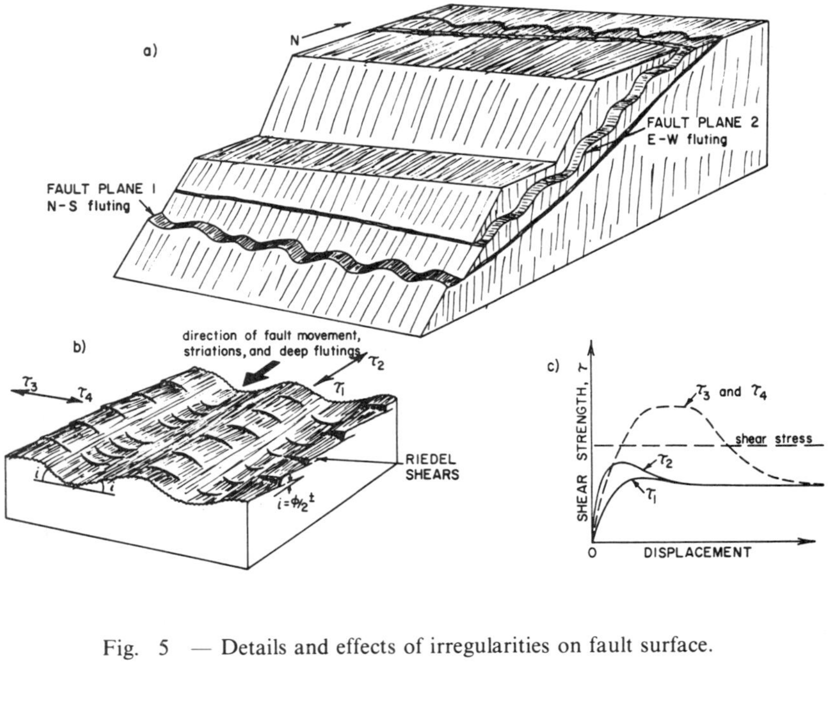

Fig. 5 — Details and effects of irregularities on fault surface.

These differences can have the effect of changing the equivalent angle of shearing resistance of a rock surface by 15° or more. Hence, the orientation of the irregularities is important in the field, in the laboratory, and in in-situ testing procedures.

The role of such oriented irregularities along a fault surface is shown in Fig. 5. Two faults are shown in Fig. 5a, both inclined with an unfavourable orientation towards the slope. The direction of movement in past geologic history along fault No. 1 was north-south resulting in flutings in this direction, whereas the direction of movement along fault No. 2 was east-west resulting in east-west oriented flutings.

Fig. 5b is a close-up sketch of a portion of one of the fault surfaces shown in Fig. 5a. The shear strength in the direction of fault movements, striations, and deep fluting is shown as τ_1 or τ_2 while the shear strength of the fault plane perpendicular to the flutings is τ_3 or τ_4.

The shear strength in the direction τ_1 may differ somewhat from τ_2 due to the presence of small Riedel shears which are sometimes observed along fault surfaces. The Riedel shears are generally observed to be inclined downward at an angle $\phi/2$ in the opposite direction to the sense of movement of the particular fault block being examined.

The value of the shearing resistance τ_3 and τ_4 in the direction perpendicular to τ_1 and τ_2 may be appreciably higher because the deep flutings must be overridden. For relatively sound rock at moderately low stress levels, the angle of shearing resistance in the direction of τ_3 or τ_4 will be the sum of the angle of shearing resistance for a flat plane plus the angle i shown on Fig. 5b.

The mechanical significance of the different orientations of the flutings is demonstrated in Fig. 5c. The shear strength-displacement curves in the direction of previous fault movement are given by curves τ_1 and τ_2, while the shear strength in the direction perpendicular is given by the curves τ_3 and τ_4. Superimposed on Fig. 5c is a horizontal line approximating the magnitude of shearing stresses (assumed to be roughly the same for both faults) acting on the two faults in the field. The position of this line with respect to the shear strength τ_1 and τ_2 indicates that fault plane No. 1 would fail, while fault plane No. 2 would not slip downhill perpendicular to the flutings because the shear strength τ_4 is greater than the magnitude of the shear stress.

Different Sizes of Irregularities

There can be several orders of magnitude of the surface irregularities. Fig. 6 is an actual trace of a bedding plane in limestone which is approximately 5 ft. long and illustrates how complex irregularities can become in nature. This figure shows a discontinuity having irregularities of two different orders of magnitude. A series of smaller second-order irregularities with larger angles of inclination from the overall dip of the surface (shown in part a) is superimposed upon a series of larger first-order irregularities (shown in part b).

Both of these sizes of irregularities can influence the shear strength of the in-situ rock mass. However, field studies of natural slopes in sandstone and carbonate rocks have indicated that the shear strength of their discontinuities is more closely associated with the smaller i-values corresponding to those obtained from the larger first-order irregularities. The implication is that some natural slope forming processes such as creep, ice formation, weathering, and cumulative displacements due to repeated application of dynamic earthquake forces tend to cause the failure of the smaller, steeper irregularities. These events leave only the larger irregularities (which require larger displacements to be overcome) to be effective in the stability of natural slopes. However, these natural geologic processes have not had time to operate in pit slopes and the minor irregularities are often responsible for the relative stability of these slopes. Other sizes of irregularities, both smaller (e.g., grain size, etc.) and larger (e.g., small flexures and folds, etc.) than those shown in Fig. 6, are significant in field and laboratory determinations of the shear strength of rock with the larger scale irregularities becoming more significant in field problems.

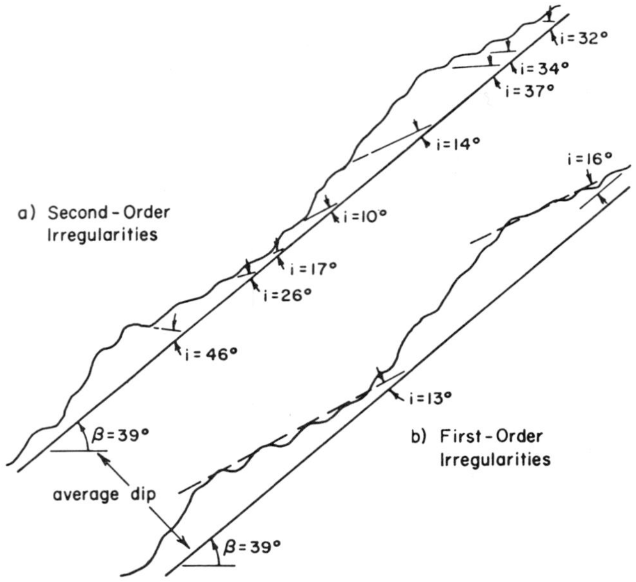

Fig. 6 — A discontinuity with first and second order irregularities -- after Patton (7).

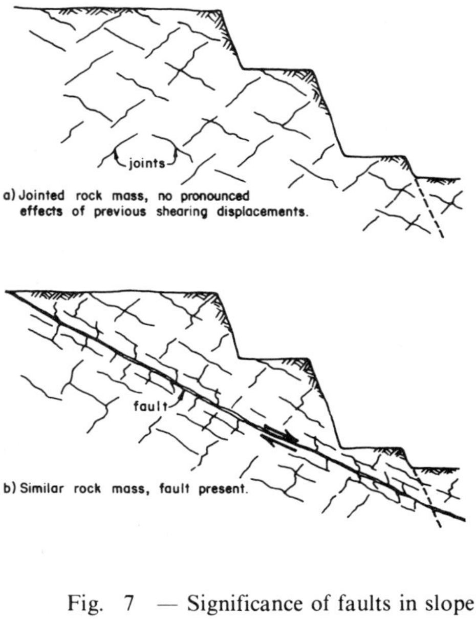

EFFECTS OF FAULTING

1. Continuity increased
   (Area of influence of structural feature is increased and cohesion, c, is reduced).
2. Irregularities reduced
   (Angle of shearing resistance, ϕ, and c are reduced).
3. Permeability altered
   (Permeability is increased, decreased, or both and unfavorable fluid pressure effects are common).
4. Weathering and alteration common
   (New, weaker minerals often present).

Fig. 7 — Significance of faults in slope stability problems.

THE IMPORTANCE OF FAULTS AND OTHER THROUGH-GOING GEOLOGIC STRUCTURES

Those experienced in the analysis of rock slopes generally recognize the much greater importance of faults and other through-going geologic structures with respect to the more plentiful discontinuous and/or irregular joints. The importance of the through-going structures may sometimes be forgotten because of the enormous amount of work and expense that may be involved in detailed joint surveys and in the plotting and analysis of these data. We believe that there are relatively few instances in which the joint orientation data turns out to be more important than knowledge of the position, orientation, and strength characteristics of the major through-going structures. A sketch summarizing many of the reasons for the increased significance of the through-going structures appears as Fig. 7.

Figure 7a shows a rock mass with discontinuous and/or irregular rock joints; Fig. 7b shows the same rock mass after a shearing displacement of natural origin has occurred along one of the pre-existing sets of joints. The effects of the shearing displacement (listed in Fig. 7) are as follows:

(a) **Continuity is increased.** Hence, the area of influence of the structure is increased and the cohesive component of the strength is decreased.

(b) **Irregularities are decreased.** Hence, both the shear strength parameters ϕ and c are reduced and the strength parameters will approach those of the residual shear strength.

(c) **Permeability is altered.** The increase or decrease in permeability can be complex. Such changes in permeability can often lead to decreases in the shearing resistance along the fault.

(d) **Weathering and alteration are common along faults.** The new weathering products are frequently clay minerals or other sheet silicates such as talc or chlorite. Hence, reduced angles of shearing resistance are common.

There are a great many names given by geologists to faults depending upon their suspected origin, sense of shearing displacement, and other characteristics. Yet, whatever their name, they tend to have many characteristics in common. Fig. 8 is a sketch of a typical cross section of a fault. The fault has a central zone of crushed and sheared rock called fault breccia (a) flanked by fine-grained often clay-rich fault gouge on either side (b) and with striated and slickensided surfaces found on the bedrock surface (c). The zone of rock (d) adjacent to either side of the fault is likely to be more highly fractured than the surrounding country rock (e). We will refer to this sequence of materials as the typical composite fault.

There are many variations to this sequence. For example, the breccia may be missing, the breccia and gouge may be missing, the fractured rock may be missing, and any or all of these layers may have been recemented. In addition, weathering often extends appreciably deeper along fault zones and hydrothermal alteration is common along faults due to the upward movement of deep-seated fluids. The weathering and alteration can superimpose additional zones of materials with different physical properties.

The most significant engineering properties of the zones in the composite fault are also indicated on Fig. 8. These include the low shear strength of the gouge-rock contact, described by Kanji (9). The shear strength of the gouge tends to be higher than what one might at first suspect as it usually contains an appreciable percentage of silt and sand-sized material. Residual angles of shearing resistance of gouge tend to be on the order of 15 to 25° but large variations are possible. For example, we have measured values of 10 to 30° for gouge containing montmorillonitic clays.

Once a fault is found and its alignments established, its influence on slope stability must be assessed. Many faults exist which have little or no influence on the stability of the

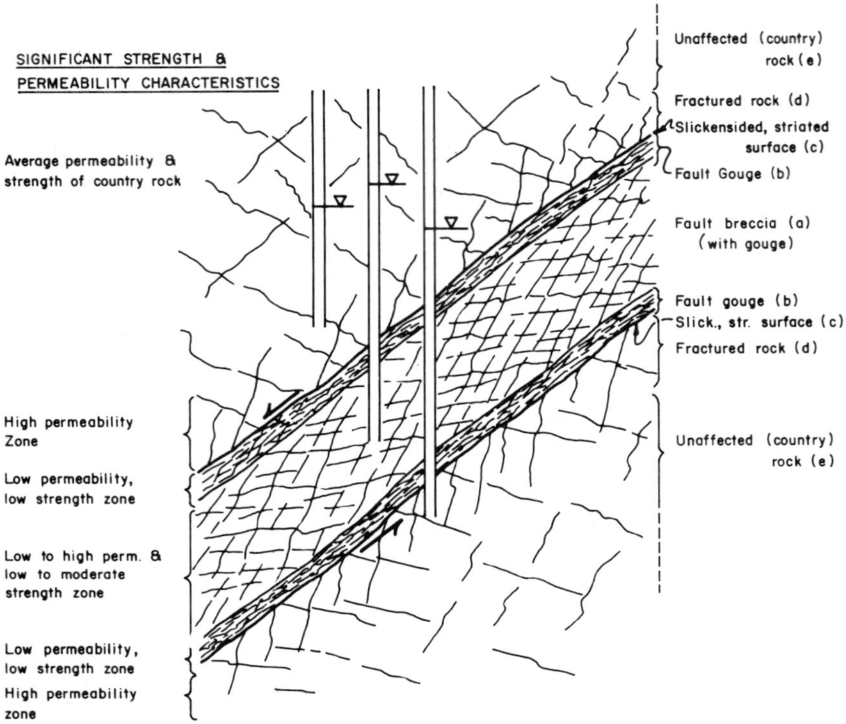

Fig. 8 — Typical cross section of a composite fault.

slopes, yet other faults and combinations of faults can be the most significant geologic factors in the analysis and prediction of the slope stability problem.

SIGNIFICANT HYDROGEOLOGIC FACTORS

Unequal or Directed Fluid Pressures

Fluid pressures within a rock mass act perpendicular to the surfaces of the discontinuities. When there are many joint sets with many different orientations and when the joint spacing is small, the fluid pressure within the rock mass can be treated in a similar fashion to that used for soil slopes. However, when the distribution of joint orientations is anisotropic and when the spacing between joints is increased, many unusual distributions of fluid pressure can result.

In rock masses it is possible to have the fluid pressures and hence the shearing forces change appreciably from one joint to the next. Fig. 9 illustrates this point for the water level is much lower in joint a-a than in joint b-b. As a result, the magnitude of the force Pb due to the hydrostatic fluid pressure along joint b-b is several times the force Pa acting normal to joint a-a. Fig. 9 also illustrates the importance and the difficulty of obtaining critical fluid pressures in rock slopes. Although we might have frequent and precise water levels recorded in piezometer 1 in joint a-a, the records are likely to be completely misleading with respect to the more critical water levels obtained from piezometer 2 in joint b-b.

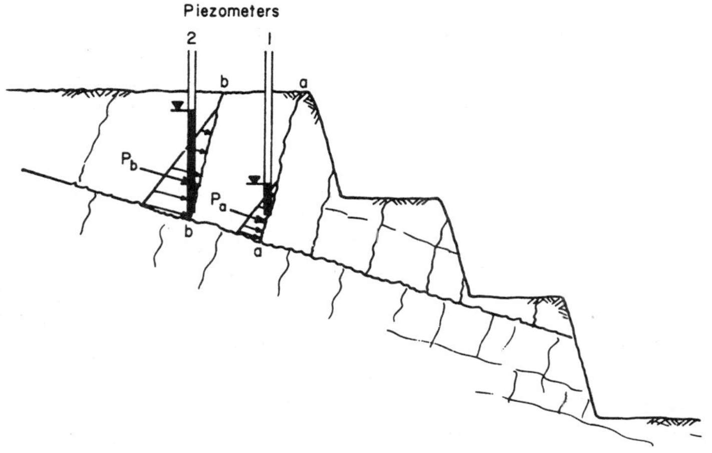

Fig. 9 — Possible large differences in fluid pressures in adjacent rock joints.

Large Fluctuations in Groundwater Levels

Groundwater levels are likely to fluctuate much more in rock slopes than in many soil slopes due to the smaller percentage of void space in rocks. Fig. 10 shows the effects on the groundwater table of a 1-in. rainfall which entirely infiltrates into a porous soil slope and a low porosity rock slope. In Fig. 10a the 1-in. rainfall can produce a 3- to 10-in. rise in the groundwater level assuming porosities of 33 to 100%, respectively. However, in Fig. 10b the same rainfall on a rock slope could produce increases in groundwater levels on the order of tens of feet. Fortunately, the rock adjacent to many rock slopes becomes more permeable because of joints opening due to blasting and stress relief. This zone of more open jointing serves to retard the development of high water pressures near the slope surface.

Fig. 10 — Comparison of groundwater fluctuations between soil and rock slopes.

Influence of Groundwater Flow Systems

The complete view of the influence of fluid pressures on the stability of mine slopes may not be available until the nature of the regional groundwater flow system is known. The principal reason for studying the regional flow system is to determine if the mine will be located in a regional groundwater recharge area, a discharge area, or in some intermediate area (see Fig. 11). The presence of a groundwater discharge area (Fig. 11a) indicates a greater possibility that excess pore-water pressures may be found in the walls and beneath the floor of the mine. In such cases the slope stability problems are likely to be aggravated. Furthermore, without some knowledge of the regional flow pattern erroneous conclusions can be drawn with respect to the suitability of drainage facilities and other remedial measures.

The position of the groundwater table often gives little indication of the portion of the regional flow system that is present. It generally takes the distribution of porewater pressure from two or more piezometers installed at different depths, such as at (a) and (b) in Fig. 11a. The increased water level in the deeper piezometer (c) from that of (d) in Fig. 11b suggests the presence of a groundwater recharge area. Although the presence of a groundwater discharge area does not necessarily mean that fluid pressures will be excessive, and the presence of a recharge area does not necessarily mean that no fluid pressure problems will be encountered, knowledge of the regional flow system is the starting point for understanding the role of fluid pressures in a mine pit.

The present authors have discussed in more detail (10), the role of groundwater flow systems in slope stability studies. In that paper the concepts developed by Hubbert (11), Tŏth (12), Freeze and Witherspoon (13) (14), and Meyboom (15) were applied to slope stability problems.

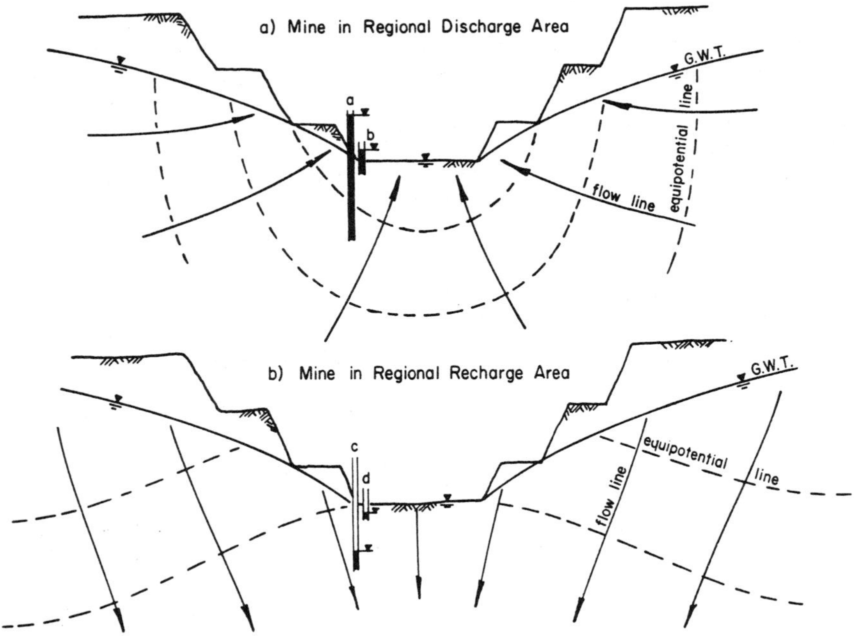

Fig. 11 — Open pit mines in different parts of a groundwater flow system.

SPECIAL EFFECTS OF FAULTS ON THE HYDROGEOLOGY OF SLOPES

A significant engineering property of faults is their effect upon the permeability of a rock mass. The typical composite fault that was shown in Fig. 8 may have one or more low permeability zones associated with the fault gouge which separates two zones of high permeability in the fractured rock. In addition, the fault breccia may be more permeable than the gouge. Thus, faults can act as groundwater barriers, as groundwater conduits, or as both at the same time. The net result of this complex layering of zones with different permeabilities is that faults can have a variety of effects on the flow of groundwater and on the resulting distributions of fluid pressures on a potential failure surface. Since the distribution of fluid pressures has a major effect on the stability of a slope, it is not uncommon to find springs and seepages of groundwater along faults which have served as a failure surface for a mine slope failure. Wilson (16) described such conditions for a failure of a portion of the Bingham Canyon pit.

Several consequences of this zonation are illustrated in Fig. 12. One possibility is that the fault may act as a groundwater barrier as shown in Fig. 12a. In this case the rock adjacent to the pit slope may be well drained yet unfavorable groundwater conditions may exist that could lead to a slope failure. Fig. 12b shows a fault serving as a groundwater conduit leading water from a nearby stream into the pit slope. In this case the dual behavior of the fault due to the presence of one of the low permeability layers in addition to the fractured rock may prevent

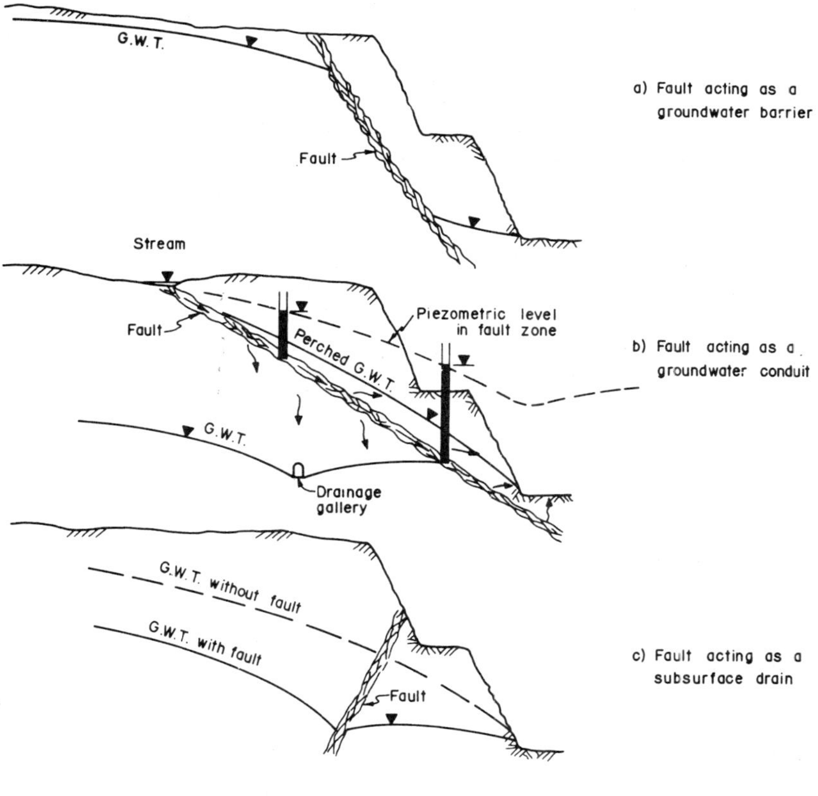

Fig. 12 — Different effects of faults on groundwater conditions.

the groundwater in the fault zone from reaching the drainage gallery shown. Fig. 12c shows a fault serving as a subsurface drain which would increase the stability of the mine slope.

It is clear that groundwater flow in and around faults and similar features, such as dykes and sills, requires special attention in slope stability studies. The physical properties of the fault zone materials must be considered as well as any change in permeability or change in physical properties due to an offsetting of lithologic units or other geologic structures. Intersections of faults require additional attention as the jointing intensity can be much higher here and the effects of weathering much deeper.

A large number of excellent papers have been written on the role of groundwater in slope stability. Included are those by Terzaghi (17) (18), Casagrande (19), Müller (20), Pacher (21), Kenney (22), Louis (23), Lane (24), Morgenstern (25), Pentz (26), Brawner (27) and Wittke (28).

It is apparent that it is more difficult to obtain representative or critical distributions of fluid pressures in rock slopes and it is more difficult to ensure that remedial drainage is effective than in most slopes in soil. Moreover, the effects of the fluid pressures may be greater than for soil slopes as the rock slopes are likely to be cut steeper.

OTHER GEOLOGIC FACTORS

Weathering and Hydrothermal Alteration

When geologic weathering occurs in soil the adverse effects are usually minor in comparison with the effects of weathering in rocks. In fact, in many instances the weathered soils are also dessicated or may be cemented and are stronger than the pre-existing unweathered soils. The effect of weathering on a rock mass is to greatly degrade its strength, to appreciably change its deformability and permeability characteristics, and to develop a complex three-dimensional arrangement of residual soil, weathered rock, and unweathered rock. In addition to being influenced by the results of surface weathering processes, mine slopes often contain decomposed rock due to hydrothermal alteration, particularly in association with faulting and igneous intrusions.

Large areas of weathered or hydrothermally altered rock may be found along wide fault zones such as that shown in the Type 3 slope failure given in Fig. 1c. Such zones often have a major influence on the groundwater flow and for this reason may cause excess pore pressures within and adjacent to the fault zone.

The influence of weathering in open pit mines is likely to be felt in the upper benches. However, in some instances, where open pit mines are cut into the sides of large hills or mountains, the stability of the upper weathered zones can appreciably influence the entire pit operation. Such a condition is illustrated in Fig. 13.

Slope failures in residual soils and weathered rocks are often closely related to relict discontinuities that remain from joints and faults in the original rock. The influence of joint irregularities is diminished as the strength of the materials forming the interior of the irregularities is reduced, thus increasing the possibility that they are sheared off rather than overridden. Weathering also tends to form zones of materials with different permeabilities which are aligned parallel to natural surfaces. The layering of lower permeability layers over higher permeability layers parallel to a slope can result in the development of artesian or excess porewater pressures in weathered slopes which may precede a slope failure. Deere and Patton (29) summarized data on typical weathering profiles, exploration, and design of slopes in residual soils.

Size Effects

In soils one does not expect appreciable changes in strength from the small, relatively undisturbed laboratory specimen to the larger in-situ test specimen to the entire failure plane whose strength is mobilized in the mine slope (excluding considerations of progressive failure).

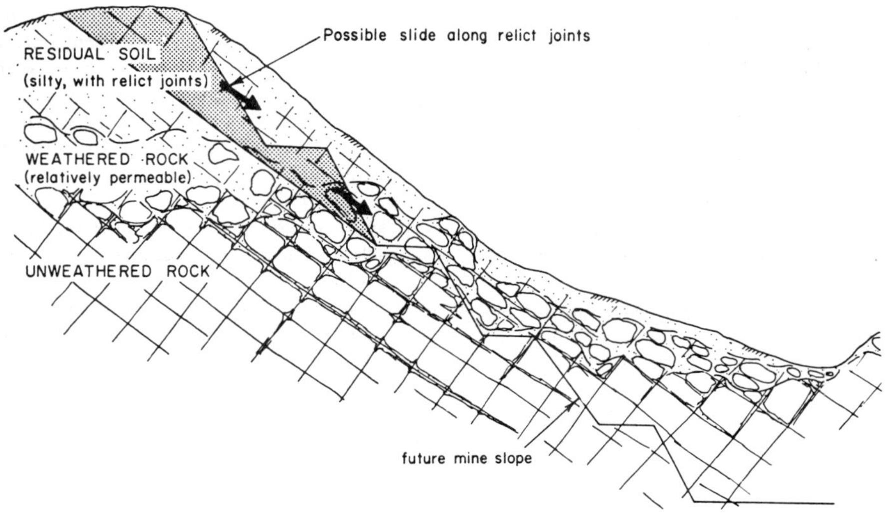

Fig. 13 — Mine slope affected by stability of residual soil and weathered rock.

However, in rock such variations in strength do occur so as to complicate the problem of correlating strengths obtained from laboratory and field tests to the overall strength that can be assumed to be mobilized along a potential failure surface in a mine slope. Such effects were described by Deere et al. (30).

Variety of Geologic Conditions Relevant to Rock Slope Stability

Many different geologic conditions can be present in local areas so as to result in (a) a variety of slope failures each with a different geologic origin, or (b) large slides in which a variety of geologic factors are operative to contribute to a complex sliding history. Whether one is required to anticipate the variety of geologic conditions that could lead to rock slides or to attempt to understand the origin of a complex rock slide, experienced field geologists are essential for obtaining meaningful conclusions from such studies. The variety of geologic details that are significant in a study of rock slopes in a given locality is likely to far exceed those encountered in soil slopes in a similar area. These points have been illustrated in a previous paper (1) with examples of rock slides from two localities — the Muskingum open pit coal mine near Zanesville, Ohio, and from highway cut slopes in eastern Peru.

Role of Regional Stresses

In considering the regional stress field, it, at first, may appear desirable to orient the shape and slopes of a mine to make tensile stresses nonexistent and to keep large concentrations of compressive stresses at a minimum. Yet, however attractive such a proposal may appear, in most practical cases the structural defects in the rock slopes are far more significant considerations for slope stability than conforming to the regional stress pattern. Only in rare instances does it seem likely that the regional stress history would appreciably influence the design of the pit layout and slopes. However, protruding horizontal or vertical corners or convex curves in the slopes should be avoided wherever possible. This is because they tend to become detached from the main rock masses due to the development of extension fractures or sheeting joints behind them.

Regional stresses can play a secondary role in pit slopes where they can cause differential movements between adjacent rock units due to stress relief strains. Such movements can have significant effects if faults with unfavourable orientations are developed in the weaker layers (see Fig. 4) and if the joints in the adjacent brittle layers open up so as to result in the development of a highly permeable zone of jointed bedrock in an adverse position. In addition, high horizontal stresses can cause the base of excavations to heave and cause severe distortions at the toe of rock slopes. However, unless unfavourable pre-existing structures are present in the pit slopes, these events should not appreciably affect slope stability.

Regional stresses may or may not be included as a causative following force acting to produce failure of a rock mass. The effectiveness of such a force would depend upon (a) the critical shear strength vs. displacement characteristics, (b) the magnitude of the displacements caused by the regional stresses, and (c) the strength required to maintain stability. In most cases it seems unlikely that displacements due to regional stresses would be sufficient to reach a strength at which the rock would fail; however, several exceptions to this generality can be imagined.

For these reasons, the principal effect of regional stresses in slope stability problems generally is to cause displacements along pre-existing planes of weakness and to some extent through limited areas of intact rock. The displacements could help to reduce the strength of the rock mass if the potential sliding surfaces had not already reached residual values of shear strength.

GENERAL TIMING OF GEOLOGICAL STUDIES

The field geology mapping and the field and laboratory testing programme will necessarily develop most fully in the early stages of the study of an open pit mine. Occasionally, borings, adits, and test pits for mineral assay purposes can be combined with requirements for developing geologic studies for slope stability. Otherwise, additional exploration work will be required. Following the documentation of the geologic framework, groundwater conditions, and physical properties, the slope stability analysis can proceed. In the following years, field mapping of new exposures should continue, tests should be performed, and new analyses made for specific problems which may appear to warrant special attention. Thus, as the mine is developed, the emphasis will gradually change from field mapping, testing and analysis to instrumentation and slope monitoring where required.

ROLE AND RESPONSIBILITY OF THE ENGINEERING GEOLOGIST

The engineering geologist in a slope stability study is usually responsible for geologic and hydrogeologic surveys, the presentation of the results of these surveys, recommendations regarding further work required, statements concerning the reliability of the data obtained, and placing the information gained in historical and physical context with the geology, hydrogeology, and physiography of the surrounding area.

The geological and hydrogeological survey should include the following items:

1. **A review of the regional geology, hydrogeology, and seismicity.** This field and office study would include a re-examination of regional structure, stratigraphy, and groundwater flow systems. Airphotos or other remote sensing techniques may be useful at this point. The review of the seismicity of the region should be included.

2. **A field study of the local subsurface geology.** Material descriptions (of both rocks and soils) are required and the local stratigraphic column must be developed if it is not already known. Adverse soil and rock types and the general sequence of permeable and impermeable zones must be established. Index property tests may be useful in detecting the rock and soil types with low shear strengths. Destructive and nondestructive testing and examination of

samples may be used. This work may include petrographic examination of thin sections for mineral content, for evidence of weathering or alteration, and for determining microscopic fabric. X-ray diffraction studies are useful to identify the types of clays or any unusual minerals which may be present. Atterberg limit tests are useful in soils, shales, and fault gouge to gain some insight into their general shear strength characteristics.

3. **A field study of the geologic structures present.** These structures would include joints, faults, and other structures that could influence the stability of the mine slopes. Two types of structures will usually be encountered — those that are discontinuous and very irregular and those that are continuous — and these must be treated separately.

(a) **Discontinuous and very irregular structures** would include many types of joints. It is sometimes useful to map the orientation, frequency or spacing, continuity, irregularity and other surface characteristics, fillings and alteration. This information may be summarized on a stereographic projection or otherwise tabulated for use in the stability analysis. When the rock mass has different systems of discontinuities or very irregular structures in different parts of the mine, it may be useful to determine the approximate boundaries of each structurally homogeneous region. Methods of collection and presentation of this orientation data have been described by Broadbent and Rippere (5), Piteau (4), Robertson (3) and by Pentz (26).

(b) **Continuous structures** would include faults, master joints, bedding plane joints, foliation joints, shear zones, mylonitic seams, dykes and sills, and unconformities. These geologic factors can be mapped and displayed on stereographic projections but it is most essential that their actual positions be mapped in the field and displayed on maps and cross sections. Other characteristics of these structures which could influence their strength and permeability should be noted and samples collected for laboratory testing. These characteristics would usually include: width of the structure; the average orientation of the structure together with the minimum and maximum dips and the "wave length" or size of rock mass that might be affected by either of the extreme dips; the presence of slickensided surfaces and striations; and the orientation of any surface linear features, gouge, breccia and other fillings, weathering and alteration products; and water seeps.

4. **A study of rock weathering and alteration.** The weathering profile should be described and delineated. The average thickness and material characteristics of each zone in the weathering profile should be noted as well as the maximum and minimum thickness and any anomalous characteristics. Any apparent relationship of springs or water seeps to the weathering profile should also be noted. Fig. 14 shows a typical weathering profile for metamorphic and intrusive igneous rocks and illustrates the terminology and proposed classification system for describing a weathering profile. Table 1 provides a key for distinguishing the zones within a weathering profile.

5. **A study of groundwater conditions.** Regional and local groundwater flow systems should be delineated in sufficient detail for realistic predictions of groundwater behavior to be made as the mine is excavated. Emphasis should be placed on locating and characterizing the groundwater aquifers or conduits, and aquicludes or barriers. The existence of different groundwater compartments should be suspected on either side of major faults, unconformities, dykes and sills, etc. Springs and surface water sinks should be noted along with flow directions. The influence or contribution of the local surface water to subsurface flow should be investigated.

Particular attention should be directed towards the possibility of low permeability layers existing below the floor of the mine or behind a proposed slope. Such low permeability layers might permit high fluid pressures to develop below or behind them and lead to heaving of the pit floor or slides in the pit slopes or both.

Special piezometers and pumping tests are frequently required in addition to water pressure tests and water level measurements that could be made in small diameter boreholes.

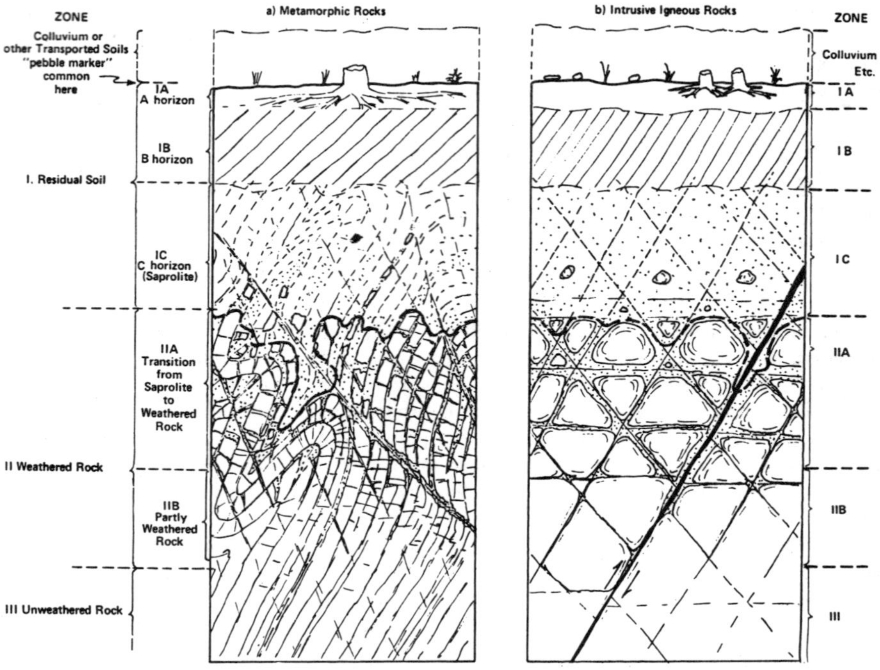

Fig. 14 — Typical weathering profile for metamorphic and intrusive igneous rocks.

The groundwater information is so important in a stability analysis that almost every borehole made in the mine area should be used to obtain groundwater data in addition to data collected for other purposes. The minimum groundwater data obtained from any boring would be the level of the groundwater table, the water level in the borehole measured every 24 hours during drilling (preferably at the beginning of each morning shift), and any unusual losses or flows of groundwater noted. An important aspect of groundwater studies in slope stability investigations is to know the magnitude of seasonal groundwater fluctuations, both in an upland area and in the vicinity of slopes. One or more permanent water level recording gauges placed early in the exploration programme will assist in estimating the magnitude of such fluctuations during mining operations.

6. **A study of existing slope failures and of the stability characteristics of local natural slopes.** Such a study is often required. In this way valuable information may be gained with a minimum of cost to the mine owner. Existing slope failures within the mine may be useful to back-calculate strength parameters and to check the results of laboratory tests and analysis. Air slaking or other surface deterioration of the existing mine slopes should be noted.

The principal problem in using past slope failures is that the magnitude of the fluid pressures acting at the time of failure can never be known. Sometimes conditions are such that the fluid pressures were probably negligible and in a few cases relatively accurate estimates can be made of the critical fluid pressures. Another problem with using old slides is that the surface or surfaces of sliding may be covered up or obliterated. Unless the geometry and strength characteristics of the surface of sliding and the fluid pressures at the time of failure

	ZONE	DESCRIPTION	RQD* (NX Core, percent)	PERCENT CORE RECOVERY* (NX Core)	RELATIVE PERMEABILITY	RELATIVE STRENGTH
I RESIDUAL SOIL	IA-A HORIZON	—top soil, roots, organic material zone of leaching and eluviation may be porous	—	O	medium to high	low to medium
	IB-B HORIZON	—characteristically clay — enriched also accumulations of Fe,Al and Si, hence may be cemented —no relict structures present	—	O	LOW	commonly LOW (high if cemented)
	IC-C HORIZON (Saprolite)	—relict rock structures retained —silty grading to sandy material —less than 10% core stones —often micaceous	O or not applicable	generally 0-10%	medium	low to medium (relict structures very significant)
II WEATHERED ROCK	IIA—TRANSITION (from residual soil or saprolite to partly weathered rock)	—highly variable, soil-like to rock-like —fines commonly fine to coarse sand (gruss) —10 to 95% core stones —spheroidal weathering common	variable, generally 0-50	variable, generally 10-90%	HIGH (water losses common)	medium to low where weak structures and relict structures are present
	IIB-PARTLY WEATHERED ROCK	—rock-like, soft to hard rock —joints stained to altered —some alteration of feldspars and micas	generally50-75%	generally>90 %	medium to high	medium to high**
III UNWEATHERED ROCK		—no iron stains to trace along joints —no weathering of feldspars and micas	>75% (generally>90%)	generally 100%	low to medium	very high**

*Notes: The descriptions provide the only reliable means of distinguishing the zones.

** Considering only intact rock masses with no adversely oriented geologic structure.

Table 1 — Description of a weathering profile for igneous and metamorphic rocks.

can be established, very little quantitative information may be gained from studies of old slides. However, even in cases where such information is lacking, valuable qualitative information can often be gathered from a study of the geologic environment of the slide.

PRESENTATION OF GEOLOGICAL DATA

The responsibility of the engineering geologist does not end with the gathering and compilation of data but must also include the presentation of the significant geologic factors in a form that is convenient, representative, and readily understandable to the slope stability analyst and to the mine management. Basically, there are two graphical methods of presenting data: (a) maps and cross sections wherein the actual positions of geological structures are shown, and (b) statistical data plots giving frequencies of orientation, spacing, continuity, etc.

1. **Maps and cross sections.** These should show the positions of all the through-going structures and major lithologic units as well as the lines of intersection of the major structures. The final versions of these plots will also show the position of the proposed engineering structures (buildings, conveyors, roads and pit slopes). The maps also include major subdivisions of the overburden materials; groundwater conditions, such as springs; and small summary statistical plots of orientation data for each structurally homogeneous region.

The preparation of maps and cross sections usually requires extrapolation or interpolation of factual geologic data. In areas where alternate interpretations are possible all possibilities should be noted. The difference between factual data and interpretations should be clearly indicated. Often, missing aspects of the field geology will not be apparent until a cross section is made, and it is not unusual for additional field inspections to be required at this time. Cross sections for stability analysis must be prepared at **true scale** or at least be accompanied by an adjacent sketch at true scale. The cross-sections should also include subdivisions of the overburden and any structures within it, the weathering profile, the groundwater table, piezometric levels and significant groundwater level fluctuations.

The main structural elements should be clearly emphasized — perhaps on separate drawings where they can be accompanied by detailed descriptions. Finally, key maps and cross sections should be included. They should show the position of the proposed mine with respect to the regional geologic and topographic features.

2. **Statistical data plots.** Field information on the orientation of planar features is commonly summarized on stereographic projections or on other graphical displays used to illustrate three-dimensional data. These displays are useful to gain some insight into the structural framework, the extent of structurally homogeneous areas, and the relative frequency of different sets of joints. The displays are especially useful when the jointing system is complex and not apparent to the casual observer.

The information may be corrected to account for the bias of the observations, particularly if the outcrops or borings available were limited in their orientations. Some of the plotting methods permit other information to be displayed, such as the continuity, irregularities, spacing, shear strength parameters, etc. Preparation of these plots can be very time consuming although they lend themselves to computerized methods of compilation and display. One problem with these displays is that a large portion of the data may never be used. Furthermore, a joint that is missed or a fault which has an orientation with a low frequency or chance of occurrence may be the truly significant feature for the stability of the mine slope. The statistical data displays draw attention to the most frequent directions of jointing but not necessarily to the most significant features.

Another use of the stereographic projection is to display data on discontinuous joints and through-going structures in a form that is readily applicable to graphical analysis. This is a useful and popular method of stability analysis but is beyond the scope of this paper (see John, (31), and Hendron et al. (32).

CONCLUSIONS

Of all the geologic factors influencing the stability of rock slopes, there is little doubt that the through-going faults and shear zones and the intersections of such structures are the most significant. Because of their continuity they can influence large areas of a pit slope and often affect more than one side of a pit. In addition, geologic displacements along faults and shear zones have led to the crushing or overriding of most irregularities in at least one direction so that low residual shear strengths are often applicable rather than the higher strengths associated with more irregular rock surfaces.

Chemical alteration of the surrounding rock and the frequent presence of breccia and clay gouge are also commonly associated with faults and shear zones. These factors lead to a decrease in the influence of surface irregularities as the intact material is more readily sheared off. Finally, the presence of clay gouge adjacent to the polished or smooth rock surfaces of faults can mean that the unusually low strengths encountered in the laboratory for soil-rock surfaces and which are developed at small displacements are applicable to the field problem. In spite of their size and continuity, the major faults and their intersections are not always readily seen until after the slope failure develops.

The groundwater conditions are also critical. Slope failures are often associated with high groundwater levels following snowmelts or intense fainfall. Faults are often associated with anomalous groundwater conditions which can be important in the stability of mine slopes.

In a paper of this length it is not possible to describe all the geologic conditions that can affect slope stability analysis, and we have only attempted to describe some of the more important factors as they appear to us in the light of our experience. We hope to have made it apparent that to find and accurately document all those geologic factors that are significant requires an extremely thorough geologic investigation. In addition, we believe we have indicated the overriding control of geology and geologic considerations in the analysis of rock slopes.

REFERENCES

1. Patton, F. D., and Deere, D. U., "Significant Geologic Factors in Rock Slope Stability", Proc. Open Pit Mining Symposium, Johannesburg, Sept., 1970.
2. Müller, L., and Hofmann, H., "Selection, Compilation and Assessment of Geological Data for the Slope Problem", Proc. Open Pit Mining Symposium, Johannesburg, Sept., 1970.
3. Robertson, A. MacG., "The Interpretation of Geological Factors for Use in Slope Theory", Proc. Open Pit Mining Symposium, Johannesburg, Sept., 1970.
4. Piteau, D. R., "Geological Factors Significant to the Stability of Slopes Cut in Rock", Proc. Open Pit Mining Symposium, Johannesburg, Sept., 1970.
5. Broadbent, C.D., and Rippere, K.H., "Fracture Studies at the Kimberley Pit", Proc. Open Pit Mining Symposium, Johannesburg, Sept., 1970.
6. Hoek, E., and Sharp, J. C., "Improving the Stability of Rock Slopes by Drainage", Proc. Open Pit Mining Symposium, Johannesburg, Sept., 1970.
7. Patton, F. D., "Multiple Modes of Shear Failure in Rock and Related Materials", Ph.D. Thesis, Dept. of Geology, Univ. of Illinois, Urbana (Univ. Microfilms Order No. 66-7786), 1966a.
8. Patton, F. D., "Multiple Modes of Shear Failure in Rock", Proc. 1st Int. Cong. Rock Mech., Vol. 1, Lisbon, 1966b, pp. 509-513.

9. Kanji, M. A., "Shear Strength of Soil-Rock Interfaces", M.S. Thesis, Dept. of Geology, Univ. of Illinois, Urbana, 1970.

10. Deere, D. U., and Patton, F. D., "Effect of Pore Pressures on Stability of Slopes", GSA-ASCE Symposium on the Significance of Pore Pressures in Problems of Engineering Geology, Annual Meeting of Geol. Soc. America, New Orleans, Nov., 1967.

11. Hubbert, M. K., "The Theory of Groundwater Motion", Jour. Geology, Vol. 48, No. 8, 1940, pp. 785-944.

12. Tōth, J., "A Theoretical Analysis of Groundwater Flow in Small Drainage Basins", Jour. Geophysical Research, Vol. 68, No. 16, 1963, pp. 4795-4812.

13. Freeze, R. A., and Witherspoon, P. A., "Theoretical Analysis of Regional Groundwater Flow: 1. Analytical and Numerical Solution to the Mathematical Model", Water Resources Research, 2, No. 4, 1966, pp. 641-656.

14. Freeze, R. A. and Witherspoon, P. A., "Theoretical Analysis of Regional Groundwater Flow: 2. Effect of Watertable Configuration and Subsurface Permeability Variation", Water Resources Research, 3, No. 2, 1967, pp. 623-634.

15. Meyboom, P., "Current Trends in Hydrogeology", Earth-Science Reviews, 2, Elsevier Publ. Co., Amsterdam, 1966, pp. 345-364.

16. Wilson, S. D., "Application of the Principles of Soil Mechanics to Open Pit Mining", Quart. Journal Colorado School of Mines, Golden, Colo. Vol. 54, No. 3, 1959.

17. Terzaghi, K., "The Mechanism of Landslides", Berkey Volume, Geol. Soc. America, 1950, pp. 83-123.

18. Terzaghi, K., "Stability of Steep Slopes in Hard Unweathered Rock", Géotechnique, Vol. 12, No. 4, 1962, pp. 251-270.

19. Casagrande, A., "Control of Seepage Through Foundations and Abutments of Dams", Géotechnique, Vol. 12, 1963, pp. 67-71.

20. Müller, L., "The Stability of Rock Bank Slopes and the Effect of Water on Same", Int. Jour. Rock Mechanics and Mining Sciences, Vol. 1, No. 4, 1964, pp. 475-504.

21. Pacher, F., "The Influence of Fissure Water on the Stability of the Rock Abutment of Arch Dams", Int. Jour. Rock Mechanics and Mining Sciences, Vol. 1, No. 3, 1964, pp. 327-339.

22. Kenney, T. C., "Stability of the Vajont Valley Slope", Rock Mechanics and Engineering Geology, Vol. 5, No. 1, 1967, pp. 10-16.

23. Louis, C., "Strömungsvorgönge in Kluftigen Medien und ihre Wirkung auf die Standsicherheit von Bauwerken and Boschungen im Fels" (Study of ground water flow in jointed rock and its influence on the stability of rock masses.) Veröffentlichungen des Institutes fur Bodenmechanik und Felsmechanik der Universität Fridericiana in Karlsruhe, Heft 30, 1967.

24. Lane, K. S., "Engineering Problems Due to Fluid Pressure in Rock", Proc. 11th Symposium on Rock Mech., Berkeley, 1969, pp. 501-540.

25. Morgenstern, N. R., "The Influence of Groundwater on Stability", Symposium on Stability for Open Pit Mining, Vancouver, 1970.

26. Pentz, D., "Methods of Evaluation and Analysis of Stability of Rock Slopes", Symposium on Stability for Open Pit Mining, Vancouver, 1970.

27. Brawner, C. O., "Case Studies of Stability on Mining Projects", Symposium on Stability for Open Pit Mining, Vancouver, 1970.

28. Wittke, W., "Three-dimensional Percolation of Fissured Rock", Proc. Open Pit Mining Symposium, Johannesburg, Sept., 1970.

29. Deere, D. U., and Patton, F. D., "Stability of Slopes in Residual Soils", State-of-the-art paper, Session II, 4th Panamer. Cong. Soil Mech. and Found. Engr., Vol. 1, San Juan, Puerto Rico, June 1971.

30. Deere, D.U., Hendron, A.J., Jr., Patton, F.D., and Cording, E.J., "Design of Surface and Near-surface Construction in Rock", Proc. 8th Symposium on Rock Mech., Amer. Inst. Mining Engrs., 1966, pp. 237-302.

31. John, K. W., "Graphical Stability Analysis of Slopes in Jointed Rock", J. Soil Mech. and Foundation Div., ASCE, Vo. 94, No. SM-2, 1968, pp. 497-526.

32. Hendron, A. J., Jr., Cording, E. J., and Aiyer, A. K., "Analytical and Graphical Methods for the Analysis of Slopes in Rock Masses", Tech. Rept. No. 36, U. S. Army Engineers, Waterways Exp. Sta., 1971.

Chapter 4

INFLUENCE OF ROCK STRUCTURE ON THE STABILITY OF ROCK SLOPES

by

Evert Hoek

Professor of Rock Mechanics,

Royal School of Mines, London, England.

SYNOPSIS

It can be shown that the properties of intact rock are relatively unimportant in rock slope stability and that failures are controlled by structural discontinuities such as bedding planes, joints and faults in the rock mass. The frequency, orientation and inclination of these discontinuities as well as their strength characteristics are important factors to be considered in any rock slope analysis and their influence will be demonstrated by means of a number of practical examples. Methods for determining the mechanical properties of discontinuities will be reviewed and the interpretation of the strength test results for use in stability computations will be discussed.

INTRODUCTION

The importance of structural discontinuities in controlling the stability of rock slopes can most conveniently be demonstrated by means of a simple practical example. Consider the case of a dry vertical slope of height H in soft limestone. Assume that the slope contains a bedding plane dipping into the excavation at an angle β as shown in Fig. 1. The uniaxial compressive strength of the intact limestone is σ_c = 288,000 lb/ft^2 (2000 lb/in^2). The strength of the bedding plane can be characterised by an angle of friction ϕ = 32° and a cohesive strength c = 2000 lb/ft^2.

When the inclination of the bedding plane is smaller than the angle of friction of this plane ($\beta < \phi$), sliding along the bedding plane cannot occur and the slope will fail as a result of failure of the intact limestone. A very crude estimate of the critical height H_c of a vertical slope in which failure occurs in this manner can be obtained by equating the uniaxial compressive strength σ_c of the material to the average vertical stress in the slope $\sigma = \frac{1}{2}\gamma\ H_c$, where γ is the density of the material. Hence

$$H_c = \frac{2\sigma_c}{\gamma} \qquad \text{........ (1)}$$

Substituting σ_c = 288,000 lb/ft. and assuming a material density γ = 160 lb/ft^3, the critical vertical height H_C = 3600 ft.

When the inclination β of the bedding plane is greater than its angle of friction ϕ = 32°, failure of the slope will occur as a result of sliding on this plane. In this case, the critical height H_C of a vertical slope is given by (1):

$$H_C = \frac{2c \sin i \cos \phi}{\gamma \sin(i - \beta) \sin(\beta - \phi)} \qquad \ldots\ldots\ldots (2)$$

where c is the cohesive strength of the bedding plane = 2000 lb/ft.^{2}

i is the slope angle = 90°

ϕ is the friction angle of the bedding plane = 32°

and γ is the rock density = 160 lb/ft.^{3}

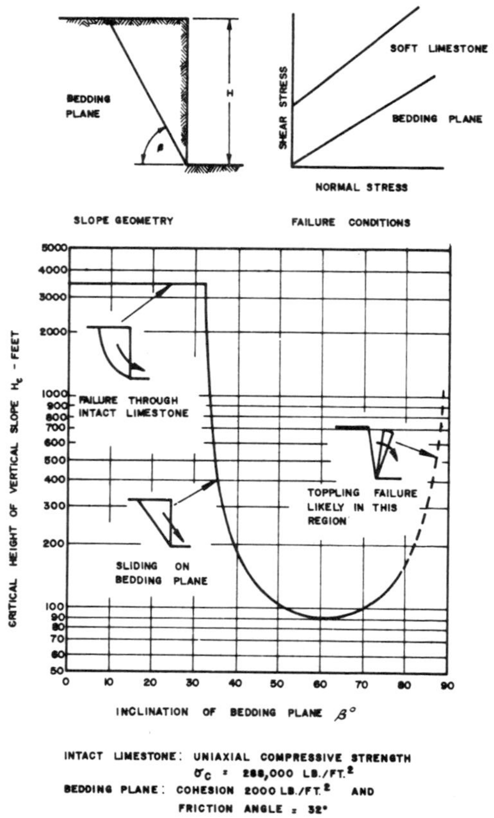

Fig. 1 — Influence of bedding plane inclination upon the critical height of a vertical dry limestone slope.

Substitution of these values and values of β between 32° and 90° gives the relationship between critical slope height H_c and bedding plane inclination β plotted in Fig. 1. Note that the most dangerous bedding plane inclination is $\beta = 61°$ and that the critical slope height for this case is 90 ft.

The fact that the presence of an unfavourably oriented structural feature such as a bedding plane, fault or joint can reduce the critical height of a vertical slope by a factor of 40 as illustrated above is ample proof of the importance of rock structure in slope stability. It is generally true to say that, in rock slopes of interest in opencast mining, failure of the intact rock would seldom occur and the stability of such slopes would be entirely controlled by the structural discontinuities within the rock mass.

STRUCTURAL FACTORS WHICH ARE IMPORTANT IN SLOPE STABILITY

Theoretical studies and practical experience of rock slope problems suggest that the structural factors which are important in slope stability are as follows:

a) The dip of joints, faults and bedding planes. As demonstrated above, this inclination plays a major role in defining the stability of a slope.

b) The strike of structural features. Obviously the relationship between the strike of important structural features and the strike of the slope face or of other structural features will define the freedom of movement of potentially unstable blocks. The most dangerous conditions occur when an unfavourably inclined structural feature strikes parallel to the slope face such that the entire face is free to slide. (Note — All two-dimensional methods of slope stability analysis assume this condition and will therefore generally err on the side of safety.)

c) The number of structural features or of sets of such features. The behaviour of a slope containing a single set of bedding planes is likely to differ from that of a slope containing three sets of intersecting joints. Not only will the entire rock mass have greater freedom to deform in the latter case but the flow of groundwater, which plays a major role in controlling the stability of a slope, will be markedly different in the two cases.

d) The continuity or persistence of structural features. In the example illustrated in Fig. 1 it was assumed that the bedding plane was continuous and possessed uniform mechanical properties over the entire slope height. Such conditions would be the exception rather than the rule in actual slopes which would normally contain a variety of geological materials and structural features.

e) The surface properties and infilling material in joints and faults. Surface characteristics such as roughness and the nature and thickness of any infilling material which may be present in a structural discontinuity have been shown to have a major influence upon the shear strength characteristics of such discontinuities (2). The permeability of these discontinuities is also significantly influenced by the presence of filling materials.

In an effort to demonstrate the influence of these various factors upon the stability of actual slopes, a number of typical two-dimensional cases are analysed and the results presented in Fig. 2. In this figure, the slope geometry and structural conditions are defined in the top row while the relevant strength values are given in the second row. In the third row, the relationships between the critical slope height (i.e. the height at which slope failure will occur) and the slope angle are given for two conditions:

a) a dry slope assuming that no groundwater is present within the potential failure region of the slope and

b) a saturated slope assuming that the groundwater surface coincides with the slope profile.

These two conditions represent the extremes of the influence of groundwater upon slope failure and the behaviour of an actual slope would normally be intermediate between these extremes.

Examination of the details presented in Fig. 2 highlights several important points in slope stability and it is convenient to consider these points in the context of specific examples.

Slopes in horizontally bedded intact rock

This case has already been considered in the example given in Fig. 1. In Fig. 2 it has been assumed that the intact material failure would follow a circular path and the slope charts for circular failure, published by the author (1), can be used to determine the critical slope height-slope angle relationship. This relationship, given in square L of Fig. 2, shows that even for the very soft limestone considered, the critical slope heights are in excess of most opencast mine slopes currently under consideration.

This confirms the conclusion reached earlier that, for slopes in materials which would normally be classed as rock, failure of the intact material is unlikely and that instability would be controlled by structural discontinuities in the rock mass.

Smooth bedding planes or joints dipping into excavation

This case has also been considered in Fig. 1 but, in Fig. 2, the influence of the potential failure plane inclination β is considered for two cases, $\beta = 45°$ and $\beta = 60°$. It will be noted that the critical slope curve becomes asymptotic to the failure plane angle β as the slope height increases. Under these conditions the wedge of material above the failure plane is held up almost entirely by friction. The cohesive strength of the discontinuity, which contributes significantly to slope stability at small slope heights, plays a decreasingly small role as the slope height increases. It will also be noted, from the graphs given in square M of Fig. 2, that the influence of groundwater pressure in the slope is more significant for the $\beta = 45°$ case than in the $\beta = 60°$ case. This difference is readily understood if the forces responsible for failure in the two cases are examined and, although such a study would exceed the scope of this paper, this fact is brought to the attention of the reader because it emphasizes the importance of defining the rock structure before embarking upon an analysis of slope stability.

Rough bedding planes or joints

Structural features such as bedding planes or joints would seldom occur as smooth continuous planes as assumed in the previous example. A more realistic analysis is based upon the assumption of a rough failure plane as illustrated in square C of Fig. 2. The strength behaviour of this type of surface would be characterised by a non-linear shear stress versus normal stress curve as given in square H of Fig. 2. At very low normal stresses, sliding along the surface occurs as a result of asperities on the rough surface riding over one another in such a way that the surfaces move apart or dilate as shear displacement takes place. As has been shown by Patton (3), this behaviour gives rise to an increased effective friction angle at low normal stress. As the normal stress increases, dilation is inhibited and sliding occurs as a result of failure of the interlocking asperities along the rough discontinuity surface. Eventually, at high normal stresses, the behaviour of such a surface could approach that of a partially cemented smooth bedding plane, the behaviour of which is indicated by the dashed line in square H of Fig. 2. This line has been purposely chosen to be the same as that in square G of Fig. 2 in order that the behaviour of slopes containing these two types of structural discontinuity may be compared.

From squares M and N in Fig. 2 it will be seen that there is very little difference between the critical slope height: slope angle curves for dry slopes with $\beta = 45°$. This is because the normal stress on the rough surface is high enough for the failure behaviour to be predominantly shearing and consequently its strength is very similar to that of the partially cemented limestone bedding plane given in square G of Fig. 2.

The introduction of water into the slope containing a rough failure plane has a dramatic

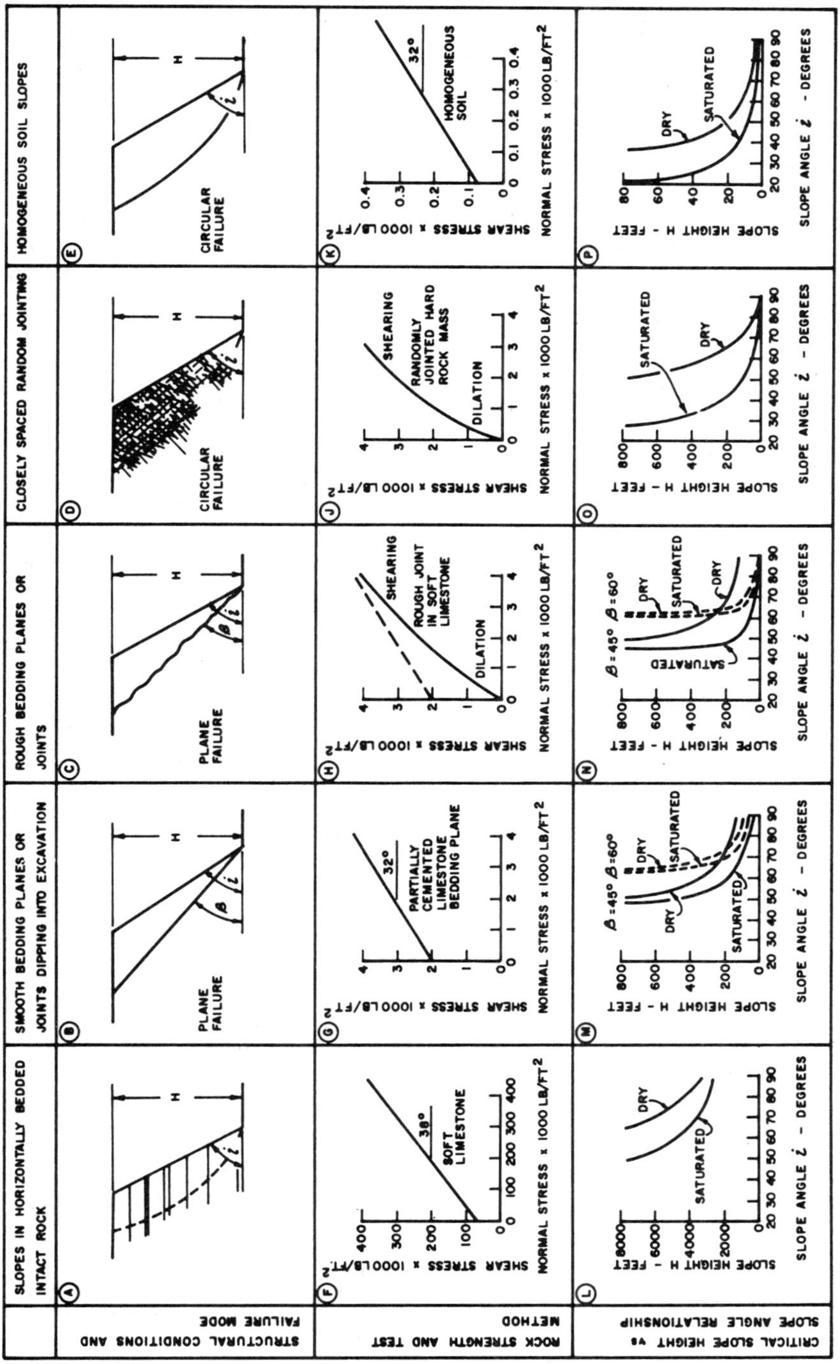

Fig. 2 —Influence of rock strength and structure upon slope stability.

effect because, in the example chosen, it reduces the normal stress to the level at which failure is associated with dilation and, although the effective friction angle of the surface is increased, its effective cohesion becomes very low. Consequently, at small slope heights where this cohesion plays an important role, the critical slope angle is significantly reduced.

The critical slope height — slope angle curves for β = 60° are both significantly different in the two cases considered because the normal stress on the failure plane is reduced, by the steepness of this plane, to a low enough level for the failure of the rough surface to be associated with dilation rather than shearing.

This particular example has been chosen to illustrate the differences between a smooth and a rough structural discontinuity. The numerical values given for slope heights and slope angles should not be taken as representative of actual situations since these would obviously vary greatly — depending upon the nature of the surface roughness. Goodman (2), has also recently shown that infilling material in a rough joint can have a significant influence upon the shear strength of that surface.

Closely spaced random jointing

In some mining situations, the rock mass may be so highly disturbed by geological processes that no systematic pattern of structural features can be discerned. When the slope height is greater than say 100 times the spacing of such randomly oriented structural features, it is not unreasonable to assume that the rock mass behaviour would be similar to that of a granular material and that the slope would fail along a circular failure surface. Jaeger (4) carried out a series of triaxial tests on undisturbed samples of a randomly jointed hard rock (Panguna andesite from the island of Bougainville) and showed that a rock mass made up of such material would have a highly non-linear failure curve such as that illustrated in square J of Fig. 2. As in the case of a rought joint surface, failure of this rock mass at low normal stresses occurs as a result of dilatant movement of the interlocking rock elements giving a very high effective friction angle. As the normal stress increases, dilation of the rock mass is inhibited and failure is associated with shearing and crushing of contact points. At very high normal stresses, when shearing of intact material is involved, the failure characteristics of the rock mass will approach those of the intact material (5).

The influence of groundwater in this slope is very important because, as in the case of the rough joint discussed in the previous section, water pressure reduces the normal stress acting across the failure surface to a level at which dilation rather than shearing takes place and the stability of the slope is seriously diminished. It will be seen that a difference of approximately 20° between the critical slope angles for dry and saturated conditions occurs in this case as compared with the average of approximately 10° given by the linear failure characteristics of the limestone (squares F and L) or the homogeneous soil (squares K and P).

Homogeneous soil slopes

This case has been included in Fig. 2 because slopes in weathered surface material, waste dumps and slimes dams on opencast mines are liable to failure by rotational movement on circular surfaces as illustrated in square E. The behaviour of this type of slope represents an extreme case in the present discussion because it is assumed that there are no significant structural features present in such a slope and hence the stability is entirely dependent upon the shear strength of the soil and upon the groundwater conditions. The stability of such slopes has been adequately discussed in soil mechanics literature (6) and does not require further consideration in this paper.

Influence of three-dimensional structure upon slope stability

The examples included in Fig. 2 are all based upon a two-dimensional analysis which assumes that the structural features upon which sliding occurs strike parallel to the slope face. Fig. 3a illustrates a typical case in which a block of material, bounded by two vertical joints normal to the slope face, slides on a structural discontinuity which dips into the excavation and strikes parallel to the face. In this case, a two-dimensional analysis will give an accurate picture of the slope failure.

A more typical situation is illustrated in Fig. 3b which shows a wedge of material sliding on two intersecting structural features. In such a case, a two dimensional analysis will not give a true picture of slope failure unless allowance is made for the distribution of forces on the two sliding surfaces (7). An approximate analysis of the stability of this wedge can be carried out by assuming two-dimensional sliding on an equivalent plane striking parallel to the slope face and dipping at the angle of the line of intersection of the two structural features. The solution obtained from such an analysis would always err on the side of safety since the shear strength of the two sliding surfaces will be higher than that of the equivalent single plane.

A more correct analysis of this type of problem has been published by John (8) and involves the use of stereographic projections to define both the structure and the force distribution within the slope. Space does not permit a complete discussion of this method and the interested reader is referred to John's work for further details.

Fig. 3c illustrates a further complication which must be considered when discussing the influence of three-dimensional structure upon slope stability — the geometry of the excavation. Since the geometry of an opencast excavation is normally defined by the shape of the ore

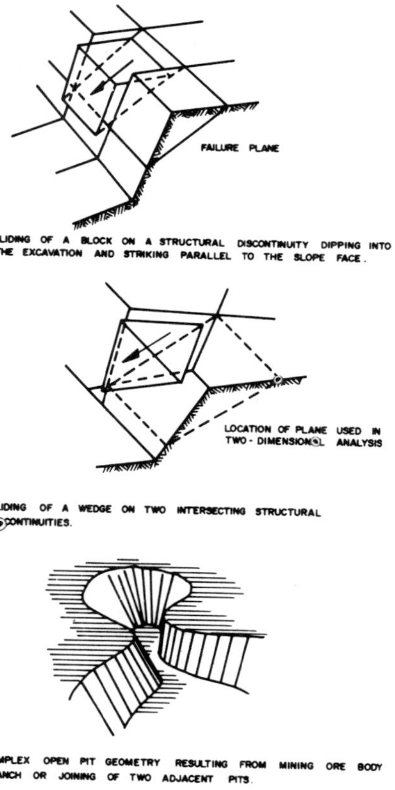

Fig. 3 —Three dimensional structures of importance in slope stability.

body being mined, the design engineer is not free to choose the most favourable pit orientation with respect to the regional structural pattern in the rock mass. Consequently, the relationship between the strike and dip of important structural features and the slope face of the excavation and the resultant influence upon slope stability will vary greatly around the pit. This variation should be kept in mind in choosing the pit slope angles and it may be found that slopes in certain portions of the pit can be appreciably steeper than the slopes in those portions in which the structural features are unfavourably oriented with respect to the slope face.

Influence of vertical joints and tension cracks upon slope stability

The examples presented in the previous sections have all been based upon the assumption that slope failure occurs as a result of sliding on a plane or circular surface which passes through the entire rock mass behind the slope face. In actual fact, this surface is usually terminated when it intersects a vertical or steeply dipping structural feature or when a tension crack is formed in the crest of the slope as illustrated in Fig. 4a. Whatever the mechanism of formation of such a tension crack, it can have a significant influence upon the stability of a slope and the detection of an open crack running parallel to the slope face can usually be taken as a

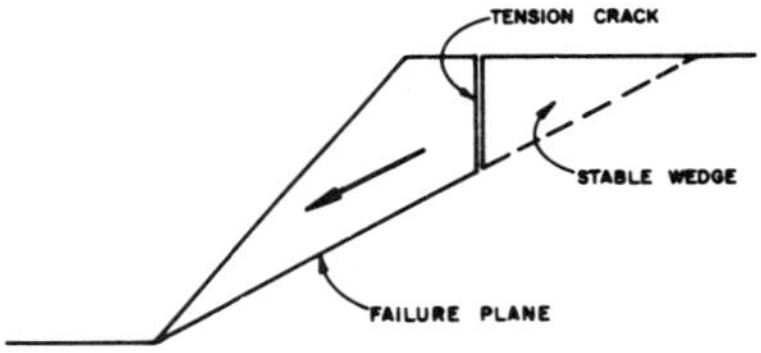

a. - TENSION CRACK IN A SLOPE.

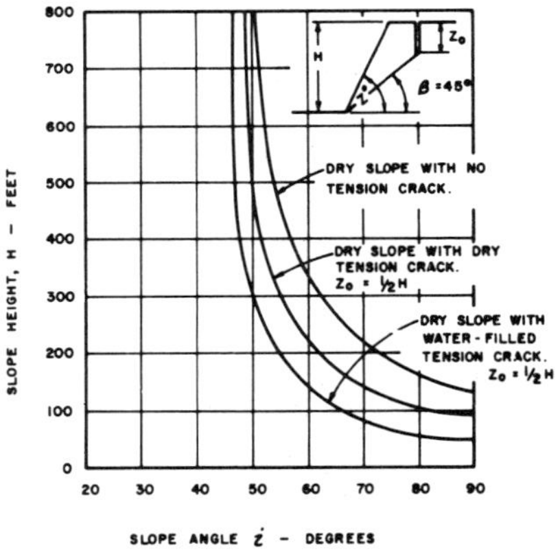

b. - CRITICAL SLOPE CURVE FOR DRY AND WATER FILLED TENSION CRACKS AND NO TENSION CRACK.

Fig. 4 —Influence of tension cracks on slope stability.

first sign of incipient slope failure. These cracks are particularly important in conditions of heavy rainfall when they can become water filled and exert a force which tends to induce slope instability.

In Fig. 4b the influence of dry and water filled tension cracks upon the critical slope height-slope angle relationship for a slope is illustrated. In this example, the same conditions as used in the analysis of plane failure in squares B, G and M of Fig. 2 were used,

i.e. $c = 2000\ lb/ft^2$, $\phi = 32°$ $\gamma = 160\ lb/ft^3$ and $\beta = 45°$

Assuming a tension crack with a depth of half the slope height ($z_0 = ½H$), the curves plotted in Fig. 4b were determined from the author's slope design charts (1). These curves show that the presence of a tension crack can have a significant influence upon the safe slope angle for a slope of given height, particularly if the crack is water filled.

Another important consequence of the presence of vertical or steeply dipping structural features in rock slopes is that toppling of columns or slabs of rock can occur. This failure mechanism is particularly dangerous in very steep slopes in which rock falls can occur as a result of toppling induced by water or ice pressure in vertical joints.

DETECTION AND MAPPING OF STRUCTURAL FEATURES

In the preceding sections of this paper an attempt has been made to demonstrate the influence of structural discontinuities such as joints, faults and bedding planes upon the stability of rock slopes. Clearly, a meaningful analysis of the stability of these slopes can only be carried out when the structural discontinuities in the rock mass have been adequately defined and the following discussion is concerned with the detection and mapping of such structural discontinuities.

Information obtained from drilling

During the early stages of development of an opencast mine, the rock mass into which the slopes of the excavation will eventually be cut may not be exposed and the only information available is that which can be obtained from diamond drill cores. The first problem facing the engineer concerned with slope design is that exploration drilling is confined to the ore body itself and there may be no boreholes in the surrounding rock. Secondly, mineral exploration drilling is normally carried out with relatively small diameter single barrel diamond drills which result in a severe disturbance of the core so that any structural interpretation is restricted to a qualitative assessment of the frequency and type of structural features which exist in the rock mass. Such qualitative information may be extremely valuable if the geologist making the interpretation has had sufficient experience of similar geological conditions and is able to deduce the most likely structural patterns which will influence the stability of future slopes.

When the stability of an opencast mine slope is important in terms of economic or operational considerations, it would normally be necessary to supplement the information available from mineral exploration drilling with drilling carried out specifically for geotechnical purposes. Such drilling would generally involve the use of relatively large diameter (up to 4½ in. diam. multiple barrel diamond drills and would require great care in core handling (9). Unfortunately there is an almost total lack of published information on this type of drilling; most of the knowledge being in the hands of drilling companies with experience in geotechnical work. Standard texts on drilling (10) tend to treat the subject somewhat superficially and the average engineer may well be unaware of the excellent core recovery which can be obtained from a carefully controlled geotechnical drilling program. It must be emphasized that this drilling is expensive as compared with normal exploration drilling and it is important that adequate financial provision should be made at an early stage in the project budget for such work.

Once the core has been recovered, there remains the problem of orienting the core with respect to the rock structure from which it was removed. One of the most detailed studies of which the author is aware, involving the reconstruction of some 800 ft. of diamond-drill core, was carried out on Mount Isa Mines in Australia by Rosengren (11). By rebuilding the entire core, Rosengren was able to work from the known orientation of structural features at the hole collar and analyse the strike and dip of all discontinuities intersected by the hole.

The expense and time involved in the type of study carried out by Rosengren has lead many engineers and geologists to consider the use of the various devices under development or commercially available for down-hole core orientation or for examination of the borehole walls with photographic, television and geophysical tools. No authoritative review of the application or potential of these methods in rock slope analysis is available but a great deal of relevant information has been published in the literature dealing with the logging of oilwells (12) (13) (14). In the author's opinion, there is much to be gained from the application of some of the geophysical tools used by the oil industry for the detection of structural features in rock masses in which slopes are to be excavated. An evaluation of such applications is currently being planned as part of the industry sponsored research project on rock slope stability being carried out at Imperial College.

Before leaving the topic of information obtainable from drilling, mention must be made of an interesting technique discussed by Dr. M. Rocha of the Laboratorio Nacional de Engenharia Civil in Lisbon at the recent Second Congress of the International Society for Rock Mechanics held in Belgrade (September 1970). Rocha described tests carried out in his laboratory in which a steel reinforcing rod was grouted into a small diameter drill hole and a reinforced column of rock was then drilled out by overcoring with a large diameter diamond drill. In this way, a detailed structural analysis could be carried out on an intact stick of core recovered from a highly fractured rock mass. Although this technique is at an early stage of development, it has many attractive features and, in the long run, its development may prove as fruitful as the development of some of the down-hole logging methods discussed earlier.

Information obtained from surface exposures

In many opencast mining situations, valuable information upon the structural characteristics of the rock mass can be obtained from adjacent opencast or underground mining excavations, exploratory adits or even from surface exposures in stream beds. In some cases, design of the final slope profiles can be deferred until some stripping has taken place and, consequently, until more information is available than that obtained from exploration drilling. In any event, it is important that adequate provision is made for the collection of structural information and this means that a geologist who has some idea of what structural features are important in controlling slope stability must be allocated the task of structural mapping.

Techniques for structural mapping vary considerably (8) (11) (15) (16) and, as is the case with most of the information gathering processes discussed in this paper, no one technique has yet been established as the optimum method for structural mapping. In the author's opinion, the actual mapping technique used is possibly less important than the level of the geologists' awareness of precisely what the design engineer requires in the way of quantitative information. At this stage in the development of rock slope stability engineering, it is probably true to say that neither the engineer nor the geologist are fully aware of what information is required and the rational development of the subject will depend, to a certain extent, upon the ability of these groups to work together on defining the problem.

As part of the research project into rock slope stability currently being carried out at Imperial College, the application of photogrammetric techniques to the mapping of structural features on exposed rock faces is being investigated. Pairs of photographs taken from accurately surveyed positions by means of a photo-theodolite are used to reconstruct a three-dimensional image from which measurements can be taken. Computer analysis of these measurements results in the definition of strike and dip of the planes in which visible structural discontinuities lie. A development of this project involves drilling holes behind a photographed face to establish whether it is permissible to project the surface planes back into the rock mass in an attempt to define the structural features which control the stability of that mass.

General remarks on structural mapping.

A critical examination of developments in slope stability engineering, as reflected in the growing literature on the subject (17), suggests that the area which is lagging furthest behind in this development is the acquisition of relevant geological data. The total lack of such data on some mines suffering slope stability problems and the expensive and time consuming techniques used to collect geological data when it is decided that such information is necessary are serious problems to the engineer required to solve slope problems. These remarks do not imply that other aspects of the subject such as theoretical methods of stability analysis, strength testing techniques, groundwater control and slope stabilization techniques have been adequately developed and do not require further attention. However, it is suggested that further development of many of these techniques without due consideration of the availability and accuracy of the geological input data is inefficient and, in some cases, irrelevant. The author is convinced that major advances in our ability to solve real slope stability problems depends, to a large extent, upon our ability to improve upon currently available techniques for structural mapping.

DETERMINATION OF THE SHEAR STRENGTH OF STRUCTURAL DISCONTINUITIES

In the early part of this paper, it was demonstrated that the stability of rock slopes is controlled by structural discontinuities in the rock mass and numerical examples were given to illustrate the influence of the shear strength characteristics of these discontinuities. The following discussion is concerned with the determination of the shear strength of structural discontinuities and with the interpretation of shear strength test results.

A review of various methods which can be used to determine the shear strength of rock surfaces has been given by Jaeger and Cook (18) and a survey of shear testing machines has recently been published by Goodman (2). Rather than repeat details contained in these publications, the principles of shear testing and some of the experimental problems involved will be illustrated by reference to a testing method currently being evaluated as part of the Imperial College rock slope stability research project. Fig. 5 shows the steps involved in a shear test carried out by means of a portable machine designed for field use. The specimen, either in the form of core or an unprepared lump containing the structural discontinuity being investigated, is mounted in quick setting plaster in such a way that the discontinuity surface lies in the shear stress plane of the machine. The shear load required to displace the discontinuity surface under a given normal load is measured and, in this machine, a total displacement of up to one inch can be accommodated. If further displacement is required in order to determine the residual or minimum shear strength of the surface, the shear and normal loads are released and the specimen set back to its original position before the test is repeated. Three or four such repetitions are normally required before the residual strength is reached. Consideration is currently being given to adding a second shear loading system so that reversal of the shear direction can be carried out without the necessity of reducing the normal load to zero and thereby risking disturbing the shear surface.

All shear testing systems suffer from similar disadvantages in that the actual distribution of shear and normal stress along the discontinuity surface is unknown (2) and the interpretation of the results can only be made in terms of average loads and contact areas. Practical difficulties such as fracture of the edges of the specimen and tipping and lateral movement of the specimen are also common to all test methods and tend to cast doubts on the absolute reliability of the results. In view of these difficulties, the author tends to view shear test results in relative rather than absolute terms and suggests that they should be regarded as indices of shear strength behaviour. Since an index requires calibration to be of any practical value, it is necessary to compare these strength measurements with actual strength values obtained by back an-

Fig. 5a —Specimen containing a structural discontinuity, set in quick setting plaster, is loaded into shear machine.

Fig. 5b —Normal and shear loads applied by means of hydraulic jacks and force required to shear specimen is measured.

Fig. 5c —Loads are released and specimen removed for examination.

alysis of slope failures.

An example of one such back analysis is presented in Fig. 6 which shows the range of peak and residual shear strengths obtained for joint surfaces in porphyry from Rio Tinto Espanola's Atalaya open pit in Spain (1). Slope angle versus slope height relationships for these strength values are plotted in Fig. 6b which also show the actual slope heights and slope angles for nine failures in porphyry slopes in mines adjacent to the Atalaya open pit. Clearly the calculated relationships, based on peak strength values, are grossly over-optimistic and, in this case, the actual slope failures are predicted reasonably accurately by a slope height-slope angle curve calculated on the basis of the average residual strength.

While it may not always be feasible to carry out an analysis of the type illustrated in Fig. 6, it is suggested that a great deal can be learned from a back analysis of existing slope failures and that this important source of data should be exploited by mining engineers who are concerned with slope problems.

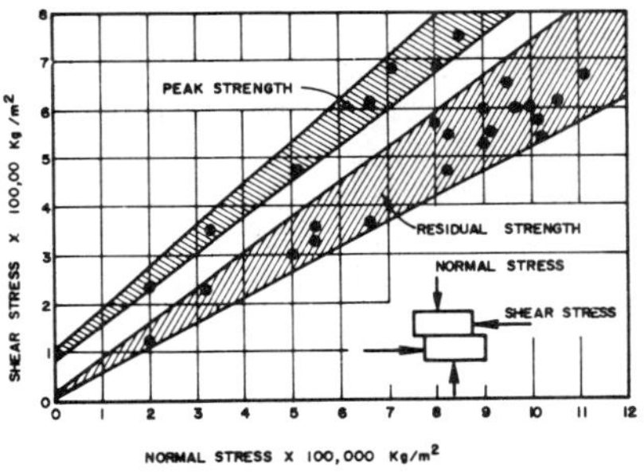

Fig. 6a —Shear strength test results for porphyry joints.

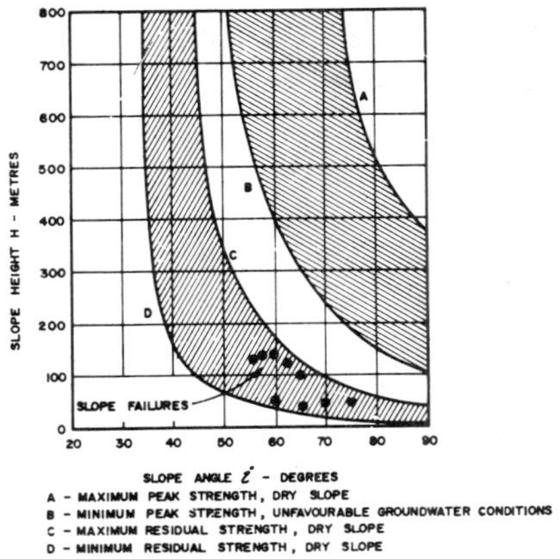

Fig. 6b —Comparison of predicted and actual porphyry slope failures.

CONCLUSIONS

The stability of slopes in materials which would normally be classed as rock is controlled by structural discontinuities such as faults, joints and bedding planes in the rock mass. Not only do these discontinuities provide surfaces upon which sliding can take place but they also control the flow of groundwater which has a very important influence on slope stability. The mapping of these structural features represents a major problem in slope stability engineering and development of mapping techniques is lagging far behind developments in other areas of slope stability analysis. Advances in our ability to design and control slopes will depend, to a large extent, upon our ability to devise more effective structural mapping techniques. It is also important that care is exercised in the interpretation of shear strength tests and it is suggested that, whenever possible, such test results be checked against actual strength characteristics as determined from the back analysis of slope failures.

ACKNOWLEDGMENTS

The work reported in this paper forms part of an industry-sponsored research project into rock slope stability in opencast mines which is being carried out under the author's direction at the Royal School of Mines, London. The project is administered by the Rio Tinto-Zinc Research Secretariat on behalf of the following companies: Anglo American International (U.K.), Ltd., Bougainville Copper Pty., Ltd., Consolidated Gold Fields, Ltd., English China Clays, Ltd., Iranian Selection Trust Ltd., National Coal Board Opencast Executive, Palabora Mining Co., Ltd., Rio Tinto Espanola, S.A., Rio Tinto-Zinc Corp., Ltd., Roan Selection Trust, Ltd., on behalf of two member companies operating in Zambia, and seven member companies of the Australian Mineral Industries Research Association, Ltd. The generous support provided by these companies is acknowledged, as is their permission to publish this paper.

REFERENCES

1. Hoek, E., "Estimating the Stability of Excavated Slopes in Opencast Mines," Trans Inst. Min. Metall (London), Section A, Volume 79, 1970, pp. A109-A132.

2. Goodman, R.E., "The Deformability of Joints", Determination of the In Situ Modulus of Deformation of Rock, ASTM, Special Tech. Publ. 477, 1970, pp. 174-196.

3. Patton, F.D., "Multiple Modes of Shear Failure in Rock and Related Materials," Ph.D. thesis, Univ. Illinois 1966, 282 p., Abstracted in Proc. 1st Congr. Intnl. Soc. Rock Mech., Vol. 1, Lisbon, 1966, pp. 509-513.

4. Jaeger, J.C., "The Behaviour of Closely Jointed Rock," Proc. 11th Symposium on Rock Mechanics, Berkeley, 1969 (in Press).

5. Rosengren, K.J. and Jaeger, J.C., "The Mechanical Properties of an Inter-locked, Low-Porosity Aggregate," Géotechnique, Vol. 18, No. 3, 1968, pp. 317-326.

6. Skempton, A.W. and Hutchinson, J., "Stability of Natural Slopes and Embankment Foundations," Proc. 7th Intnl. Conf. Soil Mech. Found. Engng., State-of-the-art report, Vol. 1, Mexico, 1969, pp. 291-340.

7. Paulding, B.W. Jr., "Coefficient of Friction of Natural Rock Surfaces," Jour. Soil Mech. Found. Div., ASCE, Vol. 96, No. SM2, 1970, pp. 385-394.

8. John, K.W., "Graphical Stability Analysis of Slopes in Jointed Rock," Jour. Soil Mech. Found. Div., ASCE, Vol. 94, No. SM2, 1968, pp. 497-526.

9. Stimpson, B., Metcalfe, R.G. and Walton, G., "A New Technique for Sealing and Packing Rock and Soil Samples," Quarterly Jour. Eng. Geol. Vol. 3, No. 3, 1970 (In Press).

10. McGregor, K., "The Drilling of Rock," C.R. Books Ltd., London, 1968.

11. Rosengren, K.J., "Rock Mechanics of the Black Star Open Cut, Mount Isa," Ph.D. Thesis, Australian National University, Canberra, 1968.

12. Schlumberger, "An Introduction to Schlumberger Well Logging," Schlumberger Document No. 8, 1958.

13. Allaud, L.A. and Ringot, J., "The High Resolution Dipmeter Tool," The Log Analyst, May-June, 1969, pp. 3-11.

14. Zemanek, J. et al, "The Borehole Televiewer: a New Logging Concept for Fracture Location and Other Types of Borehole Inspection," Jnl. Petroleum Technology, Vol. 21, 1969. pp. 762-774.

15. Robertson, A.M., "The Interpretation of Geological Factors for Use in Slope Theory," Symp. Planning Open Pit Mines, Johannesburg, 1970 (preprint).

16. Broadbent, C.D., "Fracture Studies at the Kimberley Pit," Symp. Planning Open Pit Mines, Johannesburg, 1970 (preprint).

17. Rock Mechanics Information Service, "Bibliography on Slope Stability", Rock Mechanics Abstracts, Vol. 1, No. 3, 1970, pp. 269-296.

18. Jaeger, J.C. and Cook, N.G.W., "Fundamentals of Rock Mechanics," Methuen, London, 1969.

Chapter 5

THE INFLUENCE OF GROUNDWATER ON STABILITY

by

Norbert R. Morgenstern

Professor of Civil Engineering, University of Alberta, Edmonton

INTRODUCTION

An understanding of the role of water in controlling the stability of rock masses is central for a rational approach to the subject. The presence of water can hinder mining operations in two ways:

1. excessive discharge
2. excessive pressure

It should be noted that if the rock mass is very tight, excessive pressures can exist without large discharges. Alternatively, if abundant water is available and the rock mass is open, then substantial discharges can occur under quite small pressures. The problems associated with excessive discharge are usually well understood by the mining engineer. Vehicle mobility is inhibited, some areas are flooded and hence become inaccessible; at best, a costly nuisance exists. The methods of solving these problems are also well-known to mining engineers. They generally involve tapping the source of the excessive discharge and channeling it away in such a manner that any hindrance to the mining operation is minimized. The effects of water pressure on the stability of rock masses are less obvious and the following will be devoted solely to this aspect of the problem. The manner in which water pressures enter into the consideration of stability is made clear by an understanding of the concept of effective stress.

THE CONCEPT OF EFFECTIVE STRESS

The simplest mechanical model for the problem of rock slope stability consists of a rough, rigid block sliding down an inclined plane under the influence of gravity. This is illustrated in Fig. 1. The down-slope component of the weight of the block tends to produce movement which is resisted by the shearing force generated by friction along the base of the block. The shear stress per unit area inducing movement is

$$s = \frac{W}{A} \cdot \sin \alpha \qquad \text{........ (1)}$$

while the normal stress per unit area transmitted across the slip surface is

$$p = \frac{W}{A} \cdot \cos \alpha \qquad \text{........ (2)}$$

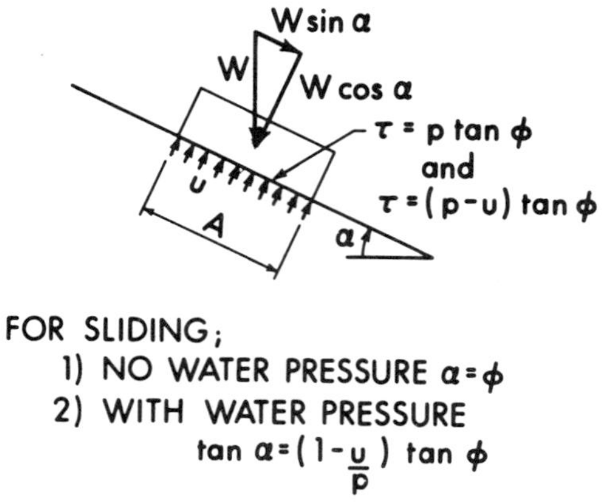

Fig. 1 —Mechanics of a sliding block.

Since the interface is rough, the normal stress induces a frictional resistance per unit area which is

$$\tau = \frac{T}{A} = p \, . \tan \phi \qquad \ldots\ldots\ldots (3)$$

where ϕ is the angle of friction at the interface. Sliding occurs when the shear stress equals the resistance, i. e.,

$$s = \tau \qquad \ldots\ldots\ldots (4)$$

$$\text{or, } \propto = \phi \qquad \ldots\ldots\ldots (5)$$

As is well-known, sliding occurs in this case when the plane is inclined to the horizontal at an angle equal to the angle of friction.

If there is a water pressure, u, acting at the interface, equation (5) is no longer true in general. The stress acting across the plane under consideration, as defined above, is called the **total** normal stress and the difference between the total stress and the water pressure, p-u, is called the **effective** normal stress. The effective normal stress will be denoted here by p^1. It was clearly recognized by Terzaghi (1)(2) that the frictional resistance to shear is controlled by the effective normal stress and not by the total stress acting across the slip surface. Hence, when water pressure effects are present, the following form of equation (3) should be used.

$$\tau = p^1 \tan \phi \qquad \ldots\ldots\ldots (6)$$

Hubbert and Rubey (3) have expounded the role of effective stresses in explaining certain geological phenomena and the significance of the concept of effective stress has been reviewed by Skempton (4).

If the equilibrium of the sliding block depicted in Fig. 1 is reconsidered in terms of effective stress when water pressure effects are present it can be shown that sliding will occur at an angle $\propto$ given by

$$\tan \propto = \frac{(1 - u)}{p} \tan \phi \qquad \ldots\ldots\ldots (7)$$

In the limit when the water pressure equals the total normal pressure the block is made fully buoyant and it develops no frictional resistance. Therefore the reduction in strength and stability of a rock mass may be substantial when water pressures in the rock mass become high.

This may be further illustrated by means of the beer-can experiment cited by Hubbert and Rubey (3). A length of plate glass is cleaned and prepared so that it will retain a continuous film of water when wetted. An empty beer-can is placed on it in an upright position and the plate is tilted until the can begins to slide. This will occur at an inclination to the horizontal of about 17°. Since the frictional resistance is independent of the area of contact, the same result will be obtained when the experiment is repeated with the can in an upside-down position. Next, the can is chilled by placing it in the freezing compartment of a refrigerator and the experiments are repeated. When the can is placed on the glass with its open end upward, the angle of sliding is found to be the same as that obtained previously. Hence the coefficient of friction of metal on wet glass is insensitive within the range of temperature of the experiment. Finally the can is rechilled and placed on the glass with its open end downward. In this case, sliding will take place at an inclination substantially less than 17°. The experiments are summarized in Fig. 2.

In the last case, the air inside the can becomes warm and as it expands the pressure increases. This increase in pressure supports the weight of the can in part. It reduces the normal component of the force between the metal and the glass without affecting the tangential component. When the can slides to the edge of the glass it will stop when the pressure is released.

A detailed explanation is illustrated in Fig. 3. If P is the normal component of the force exerted by the can on the glass, and S is the tangential or shear component, sliding will occur when

$$\frac{S}{P} = \tan \phi \qquad \dots\dots\dots (8)$$

If α is the angle of tilt, u the excess pressure of the air inside the can over that outside, A the area of the base of the can and W its weight, then

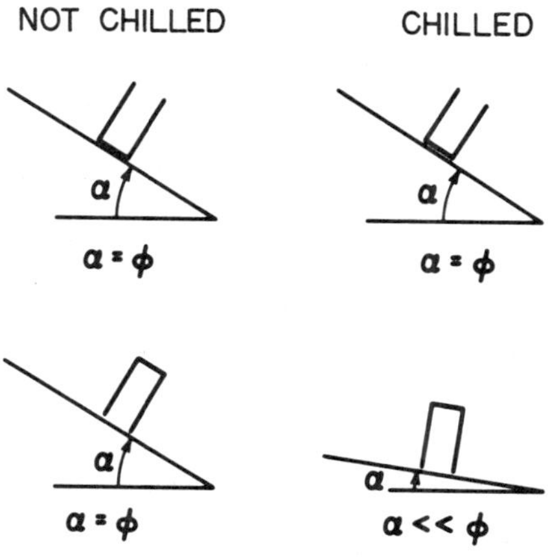

Fig. 2 —The beer can experiment.

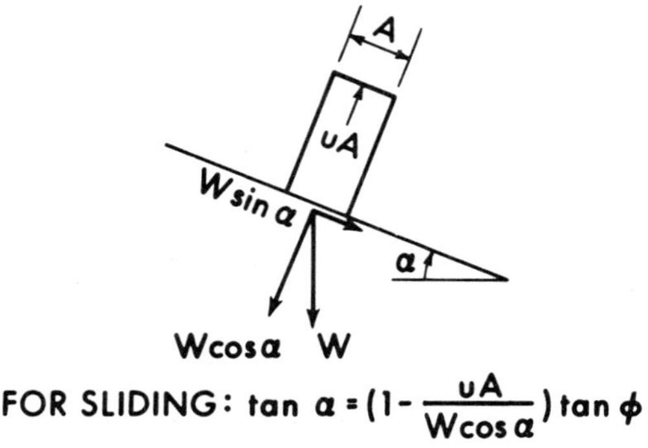

Fig. 3 —Mechanics of beer can experiment (3).

$$S = W \sin \alpha \qquad \qquad (9)$$

$$P = W \cos \alpha - uA \qquad \qquad (10)$$

If we substitute (9) and (10) into (8), it is found that sliding occurs when α satifies the following relation:

$$\tan \alpha = \left(1 - \frac{uA}{W.\cos \alpha}\right) \tan \phi \qquad \qquad (11)$$

This is the same form as equation (7) and serves to illustrate that the mechanics of this experiment and the influence of water pressures on the frictional resistance of rock are identical.

During studies associated with the solution of some landslide problems in the west of England, data were collected which reveal clearly the intimate relation between instability and changes in water pressure. (This case is drawn from the author's file). Excessive embankment loading had been placed over old landslide debris and had caused substantial movements. While remedial works were being designed observations of movements and of changes in water pressure in the debris were obtained. The slip debris was composed of broken sandstone in a sandy to silty clay matrix which was derived from the mudstone members in the intact section. The slides had been dormant until reactivation by the imposing loading. While under study during the winter of 1965-1966 more movements occurred toward the end of December, 1965 and near the end of February 1966. An example of one of the horizontal observation markers is given in Fig. 4. It shows that movements were very small until failure occurred. Then substantial deformations took place in a short period of time. They stopped almost as abruptly. (It should be noted that in general this type of behaviour will not always occur.) Unfortunately it was necessary to abandon this marker before the next series of movements occurred. Vertical displacements at a point in the active slide area are given in Fig. 5. Again large movements occurred in December 1965. Subsequently the slide stopped for several weeks but moved again in February 1966. Water pressures were measured in piezometers at several locations. One set of readings from an area immediately beside the one where displacements were being observed is given in Fig. 6. It is seen that in response to rainfall and higher rates of infiltration, water pressures rose and triggered the movements. As the pressures decreased so did the movements. But when high water pressures developed again in February, 1966 as a

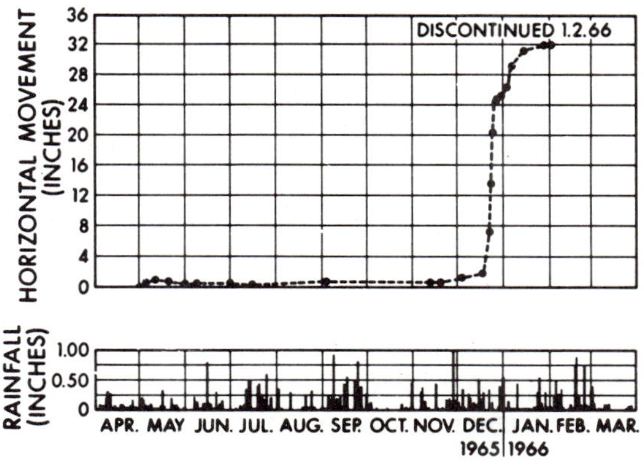

Fig. 4 —Horizontal displacement of a landslide.

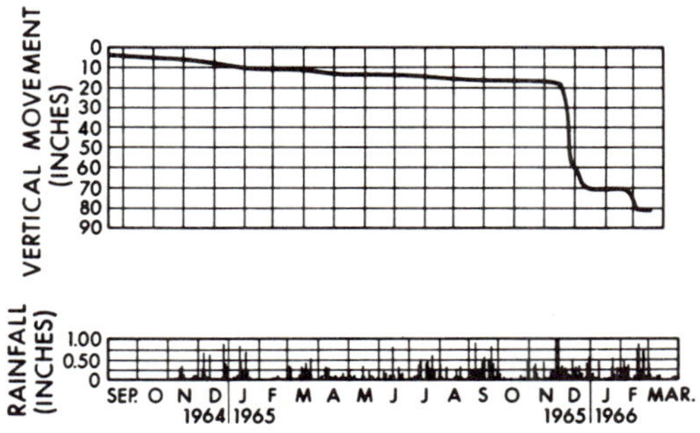

Fig. 5 —Vertical displacement of a landslide.

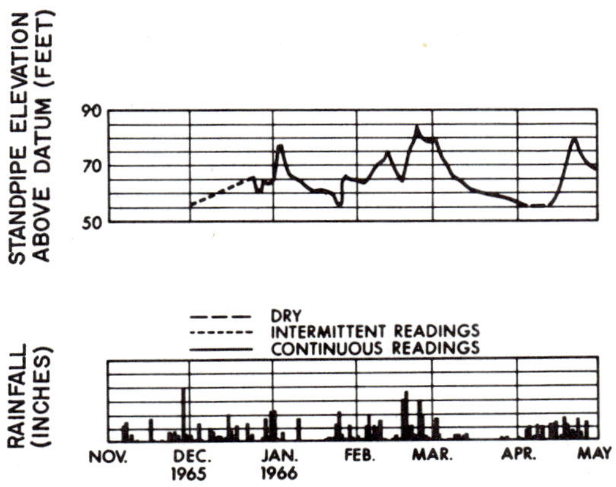

Fig. 6 —Water pressures in a landslide.

result of rainfall, movements recurred only to subside again when the pressures dissipated. This mechanism is readily understood in terms of the concept of effective stress. Another interesting feature of this example is that the slide debris was sufficiently free-draining that the water pressures dissipated readily. Had this not been the case, much larger movements would have ensued.

The determination of the water pressure distribution within a rock mass is an integral part of assessing its stability. The prediction of the change in the distribution as excavation proceeds is equally important so that the most critical conditions are anticipated with projections that are not unduly pessimistic.

PHYSICS OF FLOW OF WATER IN A ROCK MASS

In order to calculate the water pressure distribution in a rock mass and to design in-situ tests, drainage schemes and observation procedures, it is essential to understand the basic laws governing the flow of water in a porous medium. When a fluid flows through a porous medium such as sand or jointed rock a resistance to flow is generated which may be likened to the resistance to flow through a circular tube. All other things being equal, a smaller tube will offer a greater resistance to the passage of a specified discharge velocity. Alternatively, a greater pressure is required to maintain a given velocity in a smaller tube. In a similar manner a material with smaller pores and cracks will offer greater resistance to the flow of water. This behaviour is expressed quantitatively in terms of Darcy's Law.

Fig. 7 illustrates an experiment in which water flows through a column of saturated sand having a fixed length $\triangle s$. The ends of the column are connected to reservoirs whose height above some reference datum can be varied independently. Provision is made to measure the discharge Q which is the amount of water flowing through the specimen per unit time. When the discharge is measured for various differences in reservoir levels it is found that the discharge, Q, is proportional to the difference in height (h^1 - h^2). Algebraically, this may be expressed as:

$$Q = kA \frac{(h_1 - h_2).}{\triangle s} \qquad \text{........ (12)}$$

where A is the cross sectional area of the tube,

and k is a constant of proportionality called the coefficient of permeability.

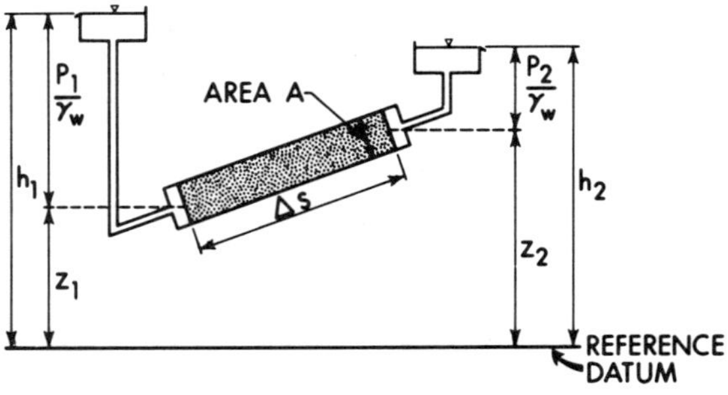

Fig. 7 —Darcy's experiment.

For a representative specimen, k will be a constant, subject to some restrictions noted in a later paragraph.

From Fig. 7, it may be seen that the height h may be expressed in terms of the pressure on the end of the column, p, and the height above the reference datum. For example,

$$h_1 = z_1 + \frac{p_n}{\gamma_w} \qquad \text{. . . . . (13)}$$

where γ_w is the bulk density of water.

Equation (13) reveals that it is both the combination of pressure and height above a reference horizontal level that causes flow. This is called the head h and is given, in general, by

$$h = z + \frac{p}{\gamma_w} \qquad \text{. . . . . (14)}$$

A physical interpretation of h may be obtained by recognizing that it is the height to which water would rise in a vertical stand pipe with one end at the point under consideration.

Equation (12) may be rewritten in terms of a discharge velocity v:

$$v = -k \frac{\triangle h}{\triangle s} \qquad \text{. . . . . (15)}$$

where $h_2 - h_1 = \triangle h$

and in the limit $\triangle S \rightarrow 0$, equation (15) becomes

$$v = -k \frac{dh}{ds} \qquad \text{. . . . . (16)}$$

Equation (16) is Darcy's Law and it expresses the direct proportionality of discharge velocity to hydraulic gradient. A comprehensive discussion of permeability has been given by Scheidogger (5).

Some typical values of permeability determined from laboratory tests on a variety of soils and rocks are assembled in Table I. It may be seen that the permeability of most rocks tested in the laboratory is extremely low. However, this is not in accord with field experience on both construction and mining projects. In numerous cases, particularly at shallow depths, the permeability has been found to be greater by several orders of magnitude. This discrepancy arises from the presence of cracks and other discontinuities in the rock mass which offer less resistance to flow than do the pores and micro-cracks in the intact rock. Fig. 8 illustrates a

TABLE I

PERMEABILITY OF SOME INTACT ROCKS

Rock Type	**Permeability** (cm/sec)
Granites	10^{-9}
Dolomite	1×10^{-8}
Limestone (1)	1×10^{-8}
Limestone (2)	1×10^{-4}
Sandstone	1×10^{-4} to 10^{-7}
Slate	1×10^{-10}

(Davis (39) has gathered a comprehensive collection of data on the porosity and permeability of natural materials.)

rock mass made of impermeable rock but containing both vertical joints and horizontal bedding planes having a separation of W and L respectively. If the pattern of discontinuities is as indicated in the figure no sample of width less than W/2 and length exceeding L can provide a continuous path of flow. Hence the measured permeability will reflect that of the intact material and will in this case be zero. However, the permeability in the ground could be appreciable. The presence of the secondary structure gives rise to a secondary permeability which for many rocks is more significant than that of the intact material, at least in the stress range of interest to civil and mining engineers.

It is of interest to investigate the discharge under laminar flow conditions through a planar crack. If the crack separation is B and the walls are smooth, the relation between fluid velocity, v_f, and hydraulic gradient is given by Lamb (6).

$$v_f = \frac{B^2}{12\mu} \cdot \gamma_w \cdot \frac{dh}{ds} \qquad \text{.........(17)}$$

where μ is the fluid viscosity.

Moreover, if there are N similar cracks in a parallel array per unit area, the discharge q per unit area will be

$$q = \frac{NB^3}{12\mu} \cdot \gamma_w \cdot \frac{dh}{ds} \qquad \text{.........(18)}$$

by analogy with Darcy's Law, the equivalent permeability of a planar array of smooth cracks is:

$$k = \frac{NB^3 \gamma_w}{12\mu} \qquad \text{.........(19)}$$

Using equation (19) it is possible to compute equivalent secondary permeabilities for combinations of N (cracks/cm) and B (cm). Fig. 9 gives some of these results. Since the model under consideration here is extremely idealized, the important conclusion to be drawn here is that it takes relatively few cracks of remarkably small aperture to increase the effective permeability of a rock mass by several orders of magnitude. However, the influence of cracks in a rock mass on the water pressure distribution may still be embraced by the concept of permeability provided that attention is given to measuring this parameter at the appropriate scale. Equation (19) also suggests that small changes in aperture openings can produce substantially larger changes in permeability.

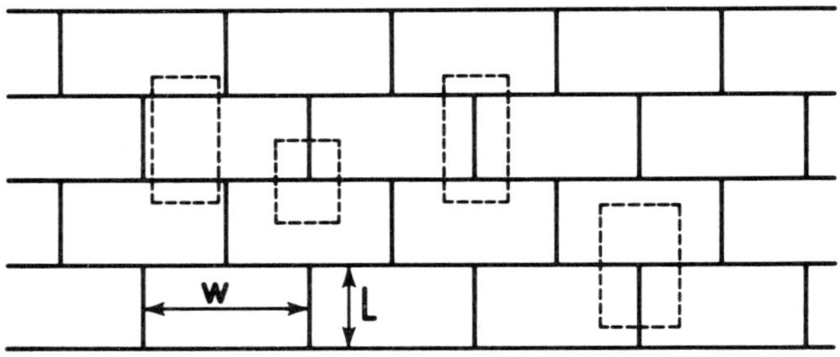

Fig. 8 —Non representative sampling (40).

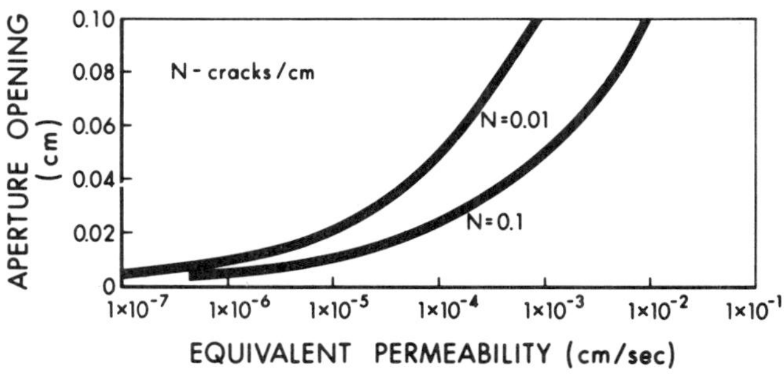

Fig. 9 —Equivalent permeability of a simple array of parallel cracks.

Equation (17) has been confirmed experimentally by Huitt (7) and Louis (8). It has been used in comprehensive theoretical studies by Snow (9). Recent studies by Sharp (10) have revealed some limitations to the validity of the equation for flows beyond the linear laminar regime and for flows in rough non-planar fissures.

For many porous materials such as gravel the preferred orientation of the particles is not marked and hence the permeability does not vary with direction. This isotropic behaviour would also be representative of a rock mass having a random distribution of joints. However, many natural materials are also anisotropic and the directional variation of permeability must be considered if the water pressure distribution is to be calculated with accuracy. Similarly, if the discontinuities in a rock mass display preferred orientations, an anisotropic variation of permeability will be needed to adequately represent its hydraulic properties. Parsons (11) has discussed the anisotropic permeability of fractured rocks. Since systematic orientation of joints and other discontinuities commonly occurs considerations of anisotropy will often arise.

In most cases of rock slope stability it is reasonable to neglect the permeability of the intact rock and consider only the secondary permeability. If the distribution of discontinuities is known in detail together with the variation of joint continuities, aperture opening, surface roughness, non-planar features, and other characteristics of discontinuities in rocks it would be possible, at least in principle, to compute the distribution of water pressure in the rock mass from elementary hydrodynamics. Except in the simplest cases, this detailed information is not available and it is preferable to determine the water pressure distribution by solving the appropriate problem of flow through an equivalent porous medium.

Prior to excavation of a slope, a pattern of flow of groundwater would have been established. Toth(12) and Freeze(13) have studied regional patterns of groundwater flow and have shown that local topography can play an important role in controlling groundwater motion and hence the water pressure distribution. As excavation proceeds the flow pattern in the area of concern will be altered. While it is true that time is required for these alterations to take place, in the following it will be assumed that for any configuration that we wish to analyze, the changes are complete and that steady-state conditions prevail. This is reasonable for all but the more impermeable materials. Procedures exist for solving transient problems (14) but they are more difficult and will not be treated here.

To complete the formulation of the problem of finding the pressure distribution it is necessary to consider the conservation of mass which, together with steady-state conditions, means that there is no net flow into or out of an element of the porous medium. Consider the element shown in Fig. 10 which for simplicity is subject to flow in only two directions. The net flow in the x direction is given by

$$\frac{\delta v_x}{\delta x} . \Delta x . \Delta y . \Delta z$$

and in the y direction by

$$\frac{\delta v_y}{\delta y} . \Delta x . \Delta y . \Delta z$$

From the condition of continuity expressed above

$$\frac{\delta v_x}{\delta x} + \frac{\delta v_y}{\delta y} = 0 \qquad \text{........ (20)}$$

If the medium is isotropic, Darcy's Law takes the following form:

$$v_x = -k \frac{dh}{dx}; \qquad v_y = -k \frac{dh}{dy} \qquad \text{........ (21)}$$

and from equations (20) and (21),

$$\frac{\delta^2 h}{\delta x^2} + \frac{\delta^2 h}{\delta y^2} = 0 \qquad \text{........ (22)}$$

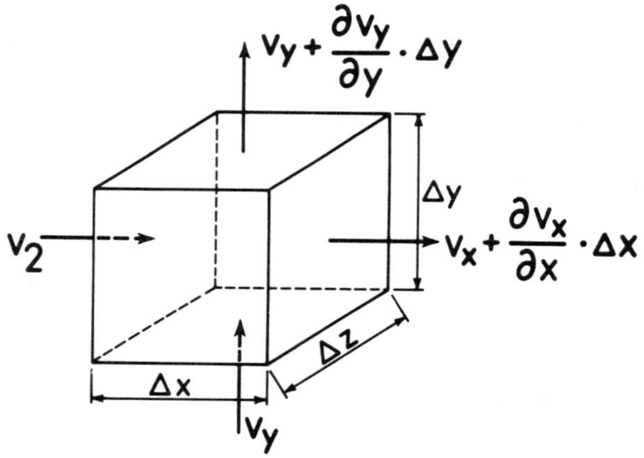

Fig. 10 —Flow into and out of an element.

Equation (22) is the well-known equation of Laplace which governs the distribution of water in an isotropic medium.

If the medium is anisotropic,

$$v_x = -k_y \frac{dh}{dx}; \qquad v_y = -k_y dh \qquad \text{........ (23)}$$

and equation (20) becomes

$$k_x \frac{\delta^2 h}{\delta x^2} + k_y \frac{\delta^2 h}{\delta y^2} = 0 \qquad \text{........ (24)}$$

Equations (22) and (24) are readily generalized to three-dimensions.

CALCULATING WATER PRESSURE DISTRIBUTIONS

In order to compute the distribution of water pressure in a rock mass it is necessary to solve the appropriate governing equation together with representative boundary conditions. The most common boundary conditions are as follows:

1. Impervious boundary where the velocity normal to the boundary must be zero since no water can cross it,
2. Constant head which is specified at a fixed location generally taken to be far enough away so as to be uninfluenced by the excavation,
3. Free surface which is the upper limit of seepage in the flow domain. The pressure at every point along it is equal to atmospheric pressure.

Solutions to the equations sufficiently accurate for most practical purposes are relatively easy to obtain, particularly for two-dimensional problems. Many excellent references are available which describe the techniques in common use. The most useful in practice are:

1. flow-net sketching
2. electrical analogue methods
3. numerical techniques using either finite difference or finite element methods.

When the available data is meagre, flow-net sketching will often provide solutions of sufficient accuracy to complement judgement in design. Anisotropy can be handled but non-homogeneity cannot be treated with ease. Meehan and Morgenstern (17) described an extremely inexpensive electrical analogue procedure which is useful for at least preliminary design work. Herbert and Rushton (18) have discussed powerful resistance network methods. When sufficient information is available to warrant the consideration of both anisotropy and non-homogeneity, the finite element method is recommended, particularly for complex geometries. (19)(20)(21).

When the water pressure distribution has been determined it is often inserted into a stability analysis which may indicate that the danger of sliding is high. One common method of increasing the resistance to sliding is to install drainage facilities and hence reduce the water pressures in the rock mass. The analysis of the influence of drainage facilities is carried out using the same procedures described here. Sharp(22) has recently studied drainage characteristics of subsurface galleries and their effect on slope stability. An example from this study illustrating the effect of a drainage gallery on lowering the water table in a 45° slope is given in Fig. 11.

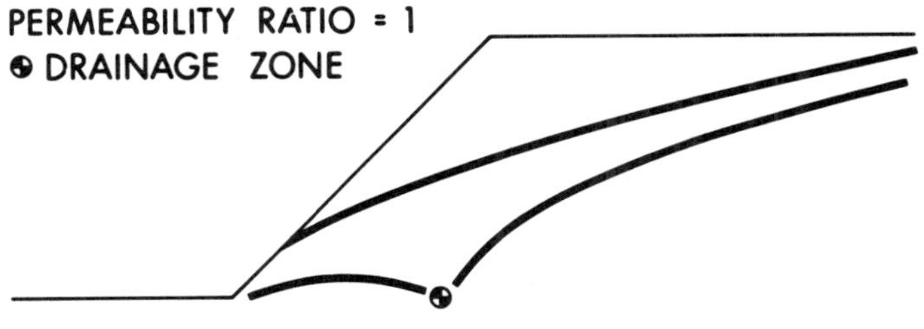

Fig. 11 —Influence of drainage gallery on water table (22).

FIELD PERMEABILITY MEASUREMENTS

If the permeability does not vary from point to point within the rock mass, a reasonable estimate of the water pressure distribution can be made without knowledge of this parameter. However it must be known for any calculations involving the amount of water flowing through the rock mass. Since different strata are usually involved in stability problems, the degree of fracturing and the continuity of fracturing may vary locally, and there may be fault zones to consider. A more realistic representation of the properties of the rock mass requires information regarding the spatial variation of the equivalent permeability. Since it is not usually practicable to take representative samples to the laboratory for testing, in-situ measurements of permeability are often carried out.

If it be assumed that at least locally the rock may be treated as an isotropic, homogeneous porous medium the equivalent permeability may be found by measuring the flow into a piezometer or packer section under a given excess head above that existing at the measuring point. This method has been suggested by Kirkham(23) and has been widely adopted. The permeability is given by

$$Q = kSh$$

where Q is the discharge in say, cc/sec.

k is the permeability in cm/sec.

h is the net head in cm.

and S is a shape factor in cm.

The shape factor S depends upon the geometry and boundary conditions in the piezometer or packed off section of borehole. Factors for most of the cavities which arise in practice have been given by Hvorslev(24) and by Al-Dhahir and Morgenstern(25). Maasland(26) has considered the influence of anisotropy on this type of measurement.

Clearly, when controlled in-situ measurements are made using piezometers or packer cavities, useful information can be obtained, particularly if the tests are carried out with due regard to observations made of the rock fabric. Tests made orthogonal to systematic discontinuities are of particular value. Moreover, if abundant tests are made, useful relative information is obtained regarding hydrogeological boundaries. While this may not be adequate for the purposes of predicting discharges such as well yields, it may suffice for the study of the pressure distribution. A stereographic projection of the discontinuities will help define the systematic orientations of the discontinuities in the rock mass and facilitate the choice of the directions of maximum and minimum flows for various planes at a point. If the principal axis of the anisotropic permeability is chosen, Snow (27) has shown how independent water pressure tests in three orthogonal drill holes along these axes may be used to determine the anisotropic variation in permeability at the point under study. While this may not be practical in many cases the analysis draws attention to the need to consider the orientation of the test with respect to the macro-fabric for useful information to be obtained.

Although they are appealing because of their simplicity, bore-hole measurements in jointed rock are not as directly relevant as in soil. Serafim(28) has drawn attention to many deficiencies, particularly when tests are carried out under high pressure. Sometimes different results will be obtained which depend upon whether the flow is increasing or decreasing. This can be a result of either cleaning out or clogging up joints with small particles. Often flow is not proportional to pressure at higher pressures. Fissure apertures will open substantially at high pressures and cause a large apparent increase in permeability. The most important deficiency of a piezometer measurement or packer test is one of scale. The spacing of discontinuities may be too large to be sampled in a representative manner and a gross underestimate of the permeability may result.

To overcome the problem of scale, pumping tests may be considered. In these tests water is pumped out of a hole and the water pressure in the hole is maintained at a specific level. Pressures are measured in adjacent boreholes away from the pumping source in order to determine the relation between pressure distribution and discharge. By using the analytical solution to the test configuration, the permeability may be computed. Solutions to many configurations in both steady and unsteady states exist, for example, Peterson (29), but those in common use are for isotropic conditions (30). Pumping tests in fractured rock have been described by Lewis and Burgy (31) and they found that the behaviour of the test was not in accord with that predicted from the theory of flow through a homogeneous, isotropic, porous medium. They attribute the discrepancies to the anisotropic and non-homogeneous characteristics of flow through a jointed rock mass.

The influence of anisotropy on pumping tests has been studied by Papadopulos (32). The anisotropy of the rock mass can be determined by observing the drawdown history in three observation wells at different distances and different directions from the pumping well. However, even these developments do not account for the data obtained by Lewis and Burgy (31). It would appear that non-homogeneity and pressure dependence of permeability are equally significant. Although pumping tests sample a large volume of rock careful consideration in their planning must be given to the rock fabric since local non-homogeneities may influence the results. Marine (33) has reported the results obtained from a variety of measurements of permeability in a fractured crystalline rock beneath the coastal plain in South Carolina. Most of the borehole tests gave a permeability less than 1×10^{-7} cm/sec, with higher values in fracture zones. However, the pumping test indicated a value many orders of magnitude higher than this. The pumping test embraced a zone of interlacing fractures of unknown extent, having a permeability that was so high that it dominated the test. Hence the result of the pumping test was not characteristic of the rock as a whole. Great care is needed in interpreting pumping tests under non-homogeneous conditions.

Permeability can vary spatially due to variations in degree and intensity of fracturing. It will also vary with depth due to the influence of increasing pressure. In some rocks this effect may be negligible; but in others it is severe. Fig. 12 illustrates data found by Bernaix (34) who tested a micro-fissured gneiss from Malpasset, France. Presumably due to the aspect ratio of the fissures, the permeability of this rock is very responsive to changes in pressure. A change of two to three orders of magnitude has been found over the pressure range of concern to most engineers. This is substantial and indicates that profound changes can take place in the permeability of the rock mass as excavation proceeds. The influence of pressure on the permeability of rock is further illustrated by the correlation between the yield of wells and their depth quoted by Davis and Turk (35) and given in Fig. 13. The correlation is based on a study of the yields of over 1500 wells in granitic and schistose rocks of the Eastern United States. The decrease of yield with increasing depth arises because the number of joints decreases with depth and they are tighter. Snow (36) has also discussed the influence of pressure changes on the change in permeability but little use has been made of these relations in practice with regard to the influence of stress release due to excavation on the water pressure distribution in a rock slope.

WATER PRESSURE MEASUREMENTS

While much is known about the flow of water through a jointed rock mass in theory, little is known in practice. This is especially true in areas of high relief. Moreover, as has been noted representative values of permeability are difficult to determine. The most reliable procedure for obtaining the distribution of water pressure in a rock mass is to measure it at a sufficient number of locations that the distribution can be interpreted with confidence. It is possible to fit a distribution of permeabilities to any set of water pressure readings which would be consistent with them and would provide a basis for predicting the influences of further excavation, pumping or drainage on the pressure distribution. Allowances can be made for the effect of pressure changes on the permeability. As excavation proceeds the pressures are monitored and either the predictions are confirmed or new input becomes available to adjust the predictor model. At the same time, the data observed in the field is directly useful in stability analyses and the design of drainage schemes. Special problems arise when sudden storms produce excessive infiltration or freezing generates an excessive buildup of water pressure by blocking the outfall. These factors cannot be incorporated in a realistic method of analysis and the best procedure to minimize the risk of water pressures causing large slides is to maintain constant surveillance of the water pressure distribution in the rock mass.

Established techniques exist for measuring and recording water pressures in rock masses. Details of the installation at the Nchanga Mine have been given by Steffen and Klingman (37). It is of interest to note that they state ". . . pore pressure observations are vital for a successful stability analysis and must be continued throughout the life of the pit, whatever the cost." One might add that the observations are only valuable if the records are kept up to date and subjected to constant interpretation. The procedure advocated here has also been adopted by Lyell (38) in his studies of the effects of water pressure on the stability of the side slopes of the Big Hole of the De Beers Mine, Kimberly.

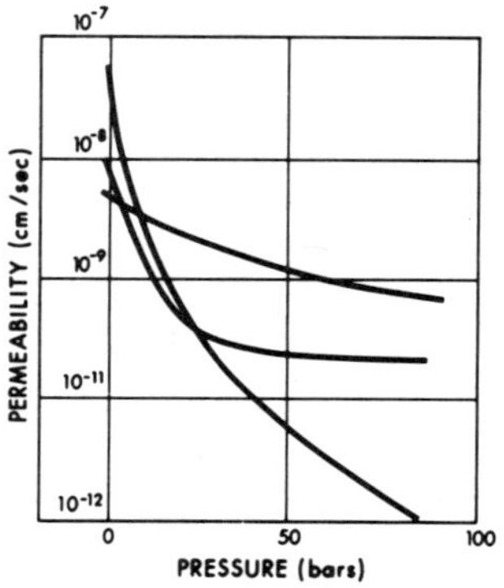

Fig. 12 — Pressure — permeability relation for rock from Malpasset (34).

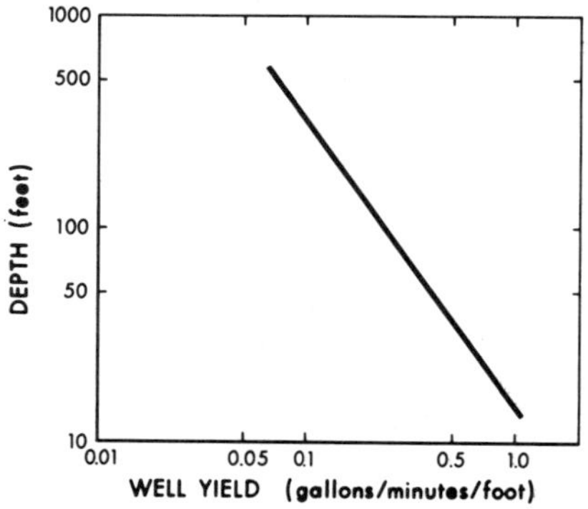

Fig. 13 — Correlation of yields in wells in crystalline rock, Eastern United States (35).

CONCLUDING REMARKS

Water pressure reduces the normal effective stress in a rock and the resistance to shear drops accordingly. Draining a rock mass will reduce the water pressures and increase the shear strength. Hence a rational stability analysis or design of a drainage stabilization scheme requires knowledge of the water pressure distribution in the rock mass.

The factors governing the flow of water through discontinuities in a rock mass have become relatively well understood in the past few years. The theory for predicting the pressure distribution is also well-developed but both the non-homogeneous and anisotropic variations of the equivalent permeability of the rock mass are needed to compute the water pressure distribution with confidence. This data is difficult to obtain. Hence it is recommended that reliance be placed on water pressures observed directly in the rock mass. This information is directly useful in a stabilitiy analysis and provides a basis for predicting the influence of further excavation or drainage on slope stability.

REFERENCES

1. Terzaghi, K., "Principles of Soil Mechanics". Engineering News Record, 95 (8 articles), 1925.
2. Terzaghi, K., "The Shearing Resistance of Saturated Soils and the Angle Between the Planes of Shear", Proc. 1st Int. Conf. Soil Mech., 1, 1936, pp. 54-56.
3. Hubbert, M.K. and Rubey, W.W.; "Role of Fluid Pressure in Mechanics of Overthrust Faulting", Bull. Geol. Soc. Amer., 70, 1959, pp. 115-116.
4. Skempton, A.W., "Significance of Terzaghi's Concept of Effective Stress", From Theory to Practice in Soil Mechanics, J. Wiley and Sons, New York, 1960.
5. Scheidogger, A.E., "The Physics of Flow Through Porous Media", Macmillan, New York, 1960.
6. Lamb., H., "Hydrodynamics" 7th ed., Cambridge University Press, London, 1959.
7. Huitt, J.L., "Fluid Flow in Simulated Fractures". J.A.I. Ch. E., 2, 1956, pp. 259-264.
8. Louis, C., "Strömungsvorgänge in kluftigen Medien and ihre Wirkung auf die Standsicherheit von Bauwerken und Böschungen in Fels". Veröff. Inst. Bodenmechanik und Felsmechanik, Universität (TH) Karlsruhe, Heft 30, 1967.
9. Snow, D.T., "Rock Fracture Spacings, Openings and Porosities," Proc. J. Soil Mechs. and Found. Div., ASCE, SMI 73-91, 1968.
10. Sharp, J.C., "Fluid Flow Through Fissured Media". Ph.D. Thesis, University of London, 1970.
11. Parsons, R.W., "Permeability of Idealized Fractured Rock". J. Soc. Pet. Eng., 6, 1966, pp. 126-136.
12. Toth, J., "A Theoretical Analysis of Groundwater Flow in Small Drainage Basins". Proc 3rd Hydrology Symposium, National Research Council of Canada, Queen's Printer, 1962, pp. 75-96.
13. Freeze, R.A., "Theoretical Analysis of Regional Groundwater Flow", Department of Energy, Mines and Resources, Inland Waters Branch, Scientific Series No. 3, 1969.
14. Maasland, D. and Bittinger, N.W., eds. "Symposium on Transient Ground Water Hydraulics". Colorado State University, Fort Collins, Colorado, 1963.
15. Harr, M.E., "Groundwater and Seepage". McGraw-Hill, New York, 1962.

16. Verruijt, A., "Theory of Groundwater Flow", Macmillan, London, 1970.

17. Meehan, R.L., and Morgenstern, N.R., "The Approximate Solution of Seepage Problems by a Simple Electrical Analogue Method", Civil Engineering and Public Works Review, 63, 1968, pp. 65-70.

18. Herbert, R. and Rushton, K.R., "Groundwater Flow Studies by Resistance Networks". Géotechnique, Vol. 16, 1966, pp. 53-75.

19. Zienkiewicz, O., Mayer, P. and Cheung, Y.K., "Solution of Anisotropic Seepage by Finite Elements". J. Eng. Mech. Div., ASCE, Vol. 92, EM1, 1966, pp. 111-120.

20. Finn, W.D.L., "Finite Element Analysis of Seepage Through Dams". Proc. J. Soil Mechs. and Found. Div., ASCE, Vol. 93, SM6, 1967, pp. 41-48.

21. Wittke, W., "Rechnerische und elektroanaloge Losung dreidimensionaler Aufgaben der Durchstromung von kluftigem Fels". Proc. 2nd Int. Cong. Rock Mechs., 3, paper 6-18, Belgrade, 1970.

22. Sharp, J.C., "Drainage Characteristics of Subsurface Galleries". Proc. 2nd Int. Cong. Rock Mechs., 3, paper 6-10, Belgrade, 1970.

23. Kirkham, D., "Proposed Method for Field Measurement of Permeability of Soil Below the Water Table". Proc. Soil Sci. Am., 10, 1946, pp. 58-68.

24. Hvorslev, M.J., "Time Lag and Soil Permeability in Ground-water Measurements", U.S. Corps of Engineers, Waterways Experiment Station, Bull. No. 36, 1951.

25. Al-Dhahir, Z.A., and Morgenstern, N.R., "Intake Factors for Cylindrical Piezometer Tips". Soil Science, 107, 1969, pp. 17-21.

26. Maasland, M., "Soil Anisotropy and Land Drainage", Drainage of Agricultural Lands, J.N. Luthin, ed., Amer. Soc. Agronomy, Madison, Wisc., 1957.

27. Snow, D.T., "Three Hole Pressure Test for Anisotropic Foundation Permeability," Rock Mechanics and Engineering Geology, 4, 1966, pp. 298-316.

28. Serafim, J.L., "Influence of Interstitial Water on the Behaviour of Rock Masses", Rock Mechanics in Engineering Practice, K. Stagg and O. Zienkiewicz, ed., John Wiley and Sons, New York, 1968, pp. 55-97.

29. Peterson, D.F., "The Theory of Drainage by Pumping from Wells", Drainage of Agricultural Lands, J.H. Luthin, ed., Amer. Soc. Agron., Madison, Wisc., 1957.

30. Theis, C.V., "Relation Between the Lowering of the Piezometric Surface and the Rate and Duration of Discharge of a Well Using Ground-water Storage", Trans. A.G.U., 16, 1935, pp. 519-524.

31. Lewis, D.C., and Burgy, R.H., "Hydraulic Characteristics of Fractured and Jointed Rock". Ground Water, 2, 1964, pp. 4-9.

32 Papadopulos, I.S., "Nonsteady Flow to a Well in an Infinite Anisotropic Aquifer", Proc. Dubrovnik Symposium on Hydrology of Fractured Rocks, AIHS - UNESCO, 1, 1967, pp. 21-31.

33. Marine, I.W., "The Permeability of Fractured Crystalline Rock at the Savannah River Plant near Aiken, South Carolina", U.S. Geol. Sur. Prof. Paper 575 — B, 203-211, 1967.

34. Bernaix, J., Etude Géotechnique de la Roche de Malpasset, Dunod, Paris, 1967.

35. Davis, S.N. and Turk, L.J., "Optimum Depth of Wells in Crystalline Rocks". Groundwater, 2, 1964, pp. 6-11.

36. Snow, D.T., "Fracture Deformation and Changes of Permeability and Storage Upon Changes of Fluid Pressure", Q. Colorado School of Mines, 63, Pt.A, 1968, pp. 201-244.

37. Steffen, O.K., and Klingman, H.L., "Slope Stability at the Open Pit of Nchanga Consolidated Copper Mines Ltd." J.S. African I.M.M., 67, 1966, pp. 140-171.

38. Lyell, K.A., "The Interpretation of Water Pressure Factors for Use in Slope Theory", Symposium on Theoretical Background to the Planning of Open Pit Mines, S. African J.M.M., Johannesburg, 1970, pp. 73-85.

39. Davis, S.N., "Porosity and Permeability of Natural Materials," Flow Through Porous Media, R. de Wiest, ed., Academic Press, London, 1969, pp. 54-89.

40. Childs, E.C., "The Physical Basis of Soil Water Phenomena". John Wiley & Sons, New York, 1969.

Chapter 6

THE INFLUENCE AND EVALUATION OF BLASTING ON STABILITY

by

Alan Bauer
and
Peter N. Calder
Dept. of Mining Engineering, Queens University, Kingston

INTRODUCTION

One of the most important decisions involved in the design of an open pit mine is the selection of a slope angle. Despite this fact, engineering procedures have developed slowly in this field. Only recently have the instruments and know-how become available to undertake a design based on sound engineering principles.

The effect of blasting on open pit slope stability is one aspect in which the state of the art has not developed fully. These effects include:

a) the effects of the ground motion associated with blast waves on slope stability, and

b) the fracturing of rock in the vicinity of slopes caused by blast waves.

Natural rock jointing plays an important role in both of these.

SUMMARY OF BLASTING EFFECTS

In recent years there has been a trend to larger sizes of blast holes in open pits to the extent that today 9⅞ in. and 12¼ in. diam. holes are the most common. The reasons for this have been economic. Larger machines are more efficient and produce at a lower overall cost per ton. As the size of blast holes have increased, so have the other factors associated with blasting. Bench heights have increased, the tons/ft. of hole yield has risen significantly and the explosive load per hole has increased. Quite naturally the rupture radius for a single blast hole has increased along with the vibration level produced per blast hole.

Open pit blasting operations can result in two types of unwanted damage:

1. Damage to structures, such as buildings, dams and slopes by ground vibrations initiated by the blast.

2. Fracturing of ground beyond the planned excavation limits.

Let us first consider item one. Vibration effects are determined by the type of blast which is fired and the type of medium through which the blast wave is being transmitted. They are strongly influenced by the weight of charge per hole, the number of holes per delay, the distance from the shot point, the mode of initiation, the point of interest relative to the blast, etc.

During the St. Lawrence Seaway project in 1957, Edwards and Northwood (1) studied the effects of blasting vibrations on building damage. Of all the variables studied they found the best correlation between damage and peak particle velocity in the disturbance. They concluded that minor damage started at peak particle velocities of 2 in./sec. in any direction. This prompted the U.S.B.M. (2) to review its records and they likewise came to the same conclusion. They also determined that major building damage occurred at particle velocities of the order of 8 in./sec. They went somewhat farther in that they developed scaling laws for particle velocity versus scaled distance of the form:

$$V = K\left(\frac{d}{W^{1/2}}\right)^{x} \qquad \text{........ (1)}$$

where

V	=	the particle velocity (in./sec.)
d	=	distance in feet from shot to observation point.
W	=	charge weight in lb. per delay.
K	=	constant depending on the distribution of the charge and the material type, varies from 45 to 450.
x	=	varies with material type and whether the longitudinal, vertical or transverse component is being measured.

As a general rule the longitudinal and vertical components are approximately equal and the transverse component is considerably smaller than the other two, so that usually the longitudinal or vertical velocities are used. The exponent x varies from -1.5 to -1.9 for these two components; a good average is -1.7. Fig. 1 shows typical peak vertical particle velocity plotted versus scaled distance in different rocks. This plot also indicates the effect in rock of distributing the charge in several holes. The variations in the data due to rock type and charge distribution often will dictate that shots be fired at the site to determine precisely the relationship between $\frac{d}{W^{1/2}}$ and particle velocity at that site. Playing it safe by keeping $\frac{d}{W^{1/2}}$ per delay less than 450 often is too conservative and makes for operational difficulties.

In rock as a general rule 9 milliseconds and longer is treated as a perfect delay interval in that no energy is assumed to feed into the peak of the waves from later delay periods. Experience that the author had with N. Farries, N. Pearson and C. Anderson (3) when blasting in tar sand indicated that this was not necessarily the case in soft material such as for sand or in overburden. The peak particle velocity in tar sand often appeared on the second cycle and about 50 milliseconds after the start of the wave. In order to stop additive effects of one charge on another it was necessary to use 100 millisecond delays. In this type of blasting, being conducted to simply "bump" the material, the charges were placed at close to the critical depth. This meant very low surface velocities that did not result in cut-offs or misfires even with the long delays used.

To investigate this further E. Progar (4) conducted a graduate student research program in which answers were sought to the effect of firing multiple charges and the effect of delays and delay interval on vibrational levels. In doing this work Progar worked with spherical charges of AN/FO of 10, 20, 30, 40 and 50 lb. in weight primed with 1 lb. pentolite primers and primacord. The tests were conducted in sandy overburden. In the first series of tests single charges of the smaller weights were fired and the vertical particle velocity was recorded at different distances from the shot point. Fig. 2 is a plot of the resulting peak vertical particle ve-

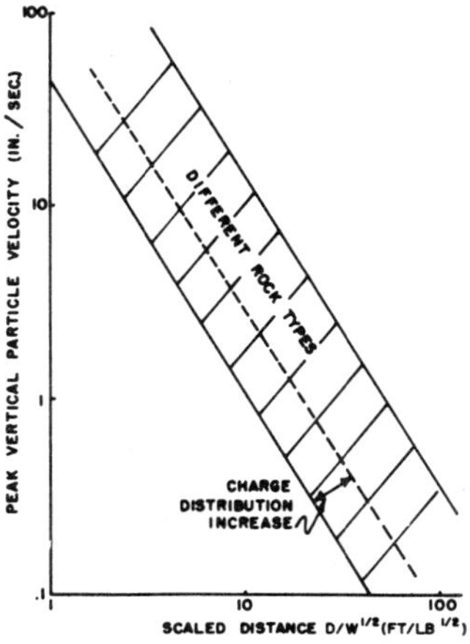

Fig. 1 —Vertical particle velocity versus scaled distance for different rocks.

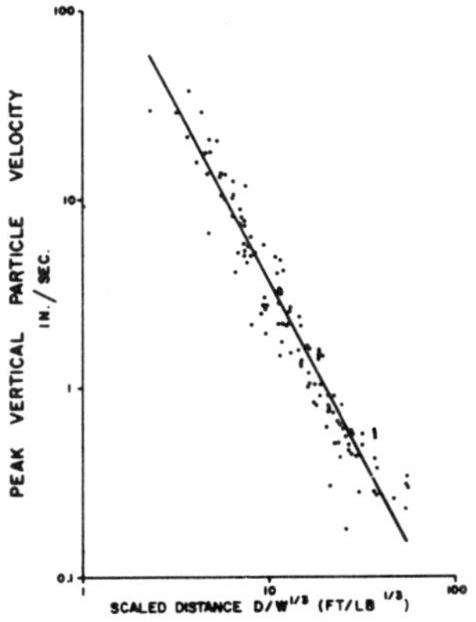

Fig. 2 —Plot of peak vertical particle velocity vs. scale distance for shots in sandy overburden.

locity versus scaled distance. Subsequent series of shots were fired employing initially 2 holes 10 ft. apart and 9 ft. deep with 30 lb. charges of AN/FO. The particle velocity was measured along the bisector and also along the line of the holes. The delays between holes were 0, 30, 50 and 75 milliseconds. Delay times were measured. No noticeable difference was detected between the 30, 50 and 75 millisecond delays, even though with 30 there should have been more reinforcement than with the other two. The subsequent work with longer rows of shots was done either instantaneously or with 30 millisecond delays between holes. Fig. 3 shows the results of the two hole shots and Fig. 4 the 7 hole shot results. The overall conclusion of this work was that for instantaneous shots the single hole peak paticle velocity data had to be multiplied by M_i and:

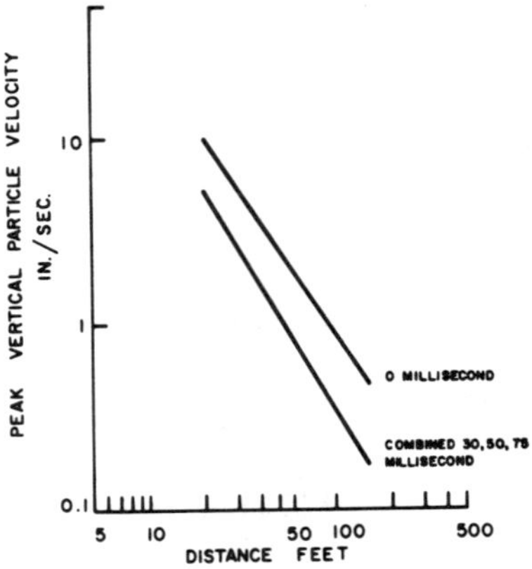

Fig. 3 —Comparison of peak vertical particle velocity vs. distance for 0 and combined 30, 50 and 75 millisecond delay 2 hole shots (gauges along the bisector).

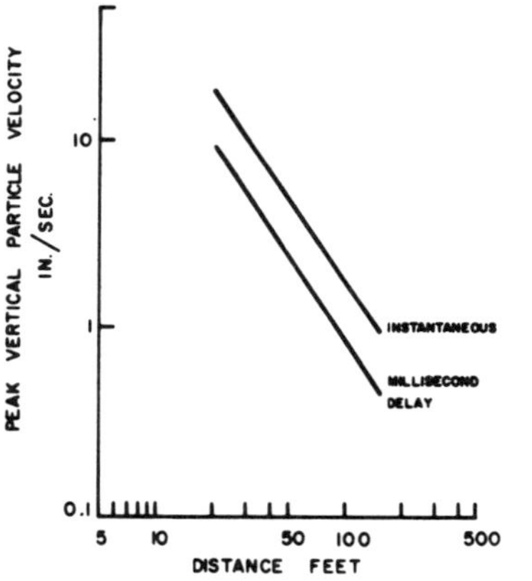

Fig. 4 —Comparison of peak vertical particle velocity vs. distance for instantaneous and millisecond delay 7 hole shots (gauges along the bisector).

$$M_i = \frac{N}{0.177\,N + 0.81} \qquad \text{........ (2)}$$

where:

N = the number of holes fired instantaneously.

For delayed shots the correction factor for particle velocity over the single hole or row value M_d is:

$$M_d = \frac{n}{0.434\,n + 1.05} \qquad \text{........ (3)}$$

where:

n = the number of delays.

For small numbers of holes per delay, up to 8, the results are essentially the same as for a charge eight times the single hole weight. For greater distribution of the charge weight the single total charge method overrates the particle velocity, for example, at 20 holes by a factor of 2.

Fig. 5 is a plot of typical ground acceleration records in rock and overburden materials vs. scaled distance. Notice that the values for rock are about two orders of magnitude greater than for overburden. Some of the implications of this will be discussed later.

The second item which has to be considered is the problem of damage to pit walls in the proximity of a blast. Cratering charges fracture rock at distances of approximately 4.5 $W^{1/3}$ or 45 ft. for a 1000 lb. charge. This is for the case of a spherical charge. Cylinders produce somewhat less extensive damage than this. Beyond this distance joints can also be opened up. Because of backbreak at the top of the bench, of the dimensions indicated above, pit operators commonly use buffer rows of holes between production blasts and final pit walls. This is referred to as control blasting and three methods are used in mining: cushion blasting, pre-shearing and buffer blasting.

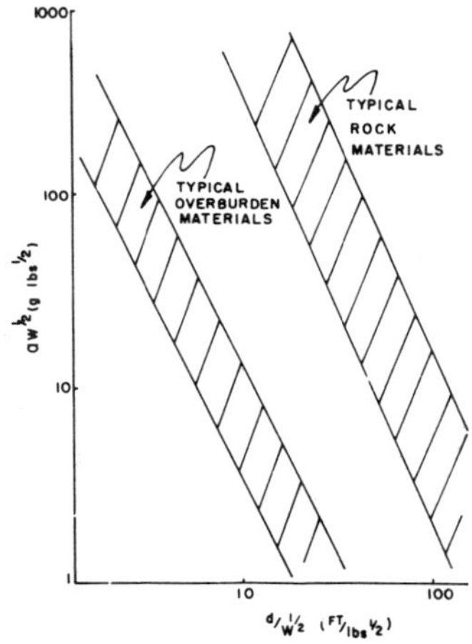

Fig. 5 —Maximum ground acceleration vs. scaled distance for typical rock and overburden materials.

Cushion blasting, sometimes referred to as slashing, involves blasting the final row of holes after the main portion of the blast has been fired and excavated. Hole loads are greatly reduced to prevent back breaking of the final wall. Hole spacing is kept somewhat less than burden to encourage the formation of a clean break between holes as in pre-shearing. Production drilling equipment can be used resulting in good economy. Fig. 6 shows a typical loaded column for cushion blasting. In this case AN/FO is being used, loaded in 4 in. cardboard tubes inserted in 9 ft. holes and pressed to the top of the hole using drill cuttings. AN/FO, having a lower detonation velocity than dynamites, produces less shattering and more "heave" from gas expansion. The shattering effect is also minimized by damping of the shock pulse passing through the uncompressed cuttings. Often the back row of the main body of the blast will be reduced in load to assist. Cushion blasting does not produce spectacular results as in the case of pre-shearing but it goes a long way in minimizing back break and the economics are favorable.

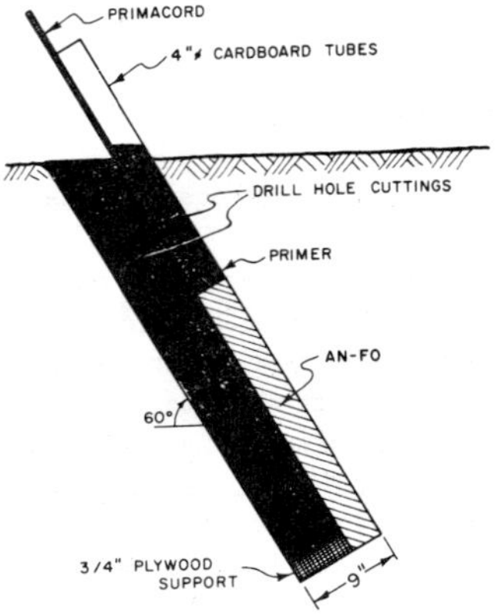

Fig. 6 —Typical loaded column for cushion blasting.

Fig. 7 shows the improvement in a section of wall that has been pre-sheared. Pre-shearing involves drilling a line of small diameter holes to conform with the desired line of break and firing these before the main body of the blast. When done properly this results in the formation of an open fracture which reflects shock waves from the main blast back into the excavation and protects the final wall from damage. When successfully carried out back break can be completely eliminated. Pre-shear holes are usually drilled with small percussion equipment, common hole sizes are in the range 2½ in. to 5 in.

A good rule of thumb for brittle rocks is that the hole spacing in feet be equal to the hole diameter in inches (i.e. 3 ft. spacing for 3 in. holes). For many mining applications satisfactory results can be obtained using hole spacings in feet up to twice the hole diameter in inches (i.e., 5 ft. spacing using 2½ in. holes). Another rule of thumb is that the explosive diameter should be half the hole diameter and the load should be distributed so that only half the length of the hole is loaded. This is usually accomplished by taping sticks of powder to the primacord downline or by stringing powder sticks with a hole along their axis on the downline and using cardboard spacers between sticks. Fig. 8 shows excellent pre-shearing results in a hard granite material typical of many Canadian open pit mine walls. Notice that the pre-shearing is at an angle to conform with the steep major joint system present. This is essential for success since vertical pre-shearing will not work where joint sets steeper than 70° are present. In this case 2½ in. holes have been used at a spacing of 4 ft. This is a much wider hole spacing than commonly used in construction practice. The results are entirely satisfactory for pit wall control. The holes were loaded with 1¼ in. x 8 in. 40% Gelatin using 8 in. spacers. The powder and spacers were strung on a line of detonator cord running the full length of the hole. Five ft. collars were used.

One often unrecognized advantage of pre-shearing is the reduction in excavation costs brought about by the elimination of overbreak. The savings effected can sometimes pay the cost of the pre-shearing. This is aside from the advantages of increased safety, savings on scaling operations and the possible justification of a steeper slope angle. One of the most important factors determining the success of a pre-shearing operation is the manner in which the final row of holes in the production blast (buffer row) is loaded. At the Adams Mine (5), for example, the spacing used in the buffer row is 20 ft. vs. 25 ft. for standard conditions. The burden between the buffer row and final row must also be reduced, usually about 25%. At Adams

Fig. 7 —Improvement in rock slope face using pre-shearing.

Fig. 8 —Angled pre-shearing conforming with major jointing.

Mine a toe load of 400 lb. of slurry explosives is used. The 10 in. hole is then loaded with a column of 5 in. stick powder to a collar of 20 ft. It is important that the pre-shear line not be left standing over a long period of time before firing the production shot. Otherwise rock chips and water will bridge the shear line and reduce its effectiveness as a reflecting boundary.

The buffer blasting technique has been developed by B. Morash (5) at Adams Mine. The concept is to distribute the explosive used in the last rows so that the energy is not so concentrated. Regular production holes are used at reduced spacing, usually 16 ft. for 10 in. holes. A toe load of 150 lb. slurry topped off with a 5 in. column of stick powder to a 20 ft. collar is used. The back row is fired simultaneously on the last delay. The method is similar in many ways to cushion blasting except that it is not necessary to excavate the main blast first. The method gives satisfactory results in fairly competent ground and may evolve into a practical economic solution to the wall control blasting problem in open pits.

THE EFFECT OF GROUND MOTIONS ON SLOPE STABILITY COMPUTATIONS

The main force which tends to produce sliding of a rock mass is that component of its own weight which acts in the direction of the steepest dip component along which sliding is possible. The normal force which is responsible for generating the frictional resistance to sliding is also a function of the weight of the rock mass. The weight of a rock block of course is proportional to its mass and acceleration. Suppose, however, the acceleration field changes and the block is no longer acted on simply by the earth's vertical pull. The entire stability force system changes dramatically. This is what happens to a rock mass during the superimposition of a blast-acceleration pulse on the earth's gravity field.

Consider the simple case of a block resting on a plane and free to slide, the factor of safety is given by:

$$F = \frac{W \cos \beta \tan \phi + C A}{W \sin \beta} \quad \ldots\ldots\ldots (4)$$

where:

W = weight of block

β = the dip angle of the slide plane

ϕ = angle of friction between block and plane

C = cohesion between the block and plane

A = area of contact

The forces acting on the rock mass are as shown in Fig. 9A. If the block were acted on by an acceleration pulse of uniform magnitude, with a component in the horizontal plane perpendicular to the face equal to a_H, with no vertical component, the force system would be as shown in Fig. 9B. The accelerated weight (W_A) would be:

$$W_A = W/\cos \theta_A \quad \ldots\ldots\ldots (5)$$

where: $\theta_A = \tan^{-1} a_H/g$

g = acceleration due to gravity.

In this acceleration field, the driving force is $W_A \sin(\beta + \theta_A)$ and the normal force $W_A \cos(\beta + \theta_A)$. (see Fig. 10). The factor of safety becomes:

$$F = \frac{W_A (\cos \beta + \theta_A) \tan \phi + C A}{W_A \sin(\beta + \theta_A)} \quad \ldots\ldots\ldots (6)$$

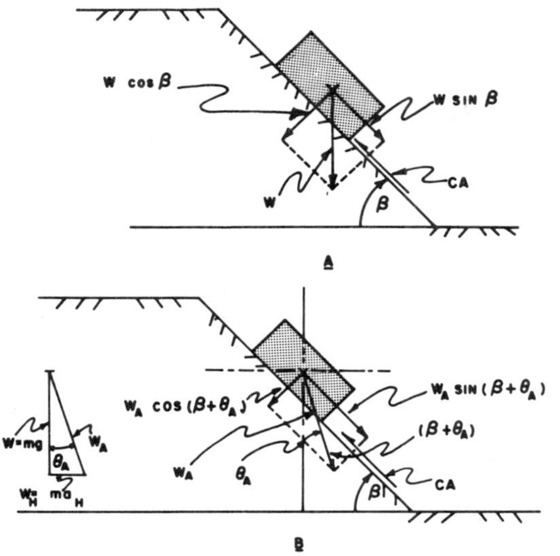

Fig. 9. —Forces acting on a rock mass, (A) in a gravity field and (B) in a gravity and blast acceleration field.

Note that for the case of zero cohesion as is often assumed in practice:

$$F = \text{Tan}\,\phi / \tan(\beta + \theta_A) \qquad \text{........ (7)}$$

Thus any change in the direction of the weight vector from the vertical is equivalent to a similar change in the slope angle.

A simple stability analysis for an overburden slope will now be considered to demonstrate the effect of blasting on these types of slopes. Consider the toe slip circle in Fig. 10A. The normal procedure is to divide the slope into a number of slices as shown. The length of arc L through the slope is easily determined from the radius (r) and subtended angle γ in radians as $L = r\gamma$. Ignoring side effects between wedges and assuming there are no water pressures present the factor of safety can be estimated for the trial failure circle as follows:

$$F = \frac{C\,L + \Sigma\,W \cos\theta_b \tan\phi}{\Sigma\,W \sin\theta_b} \qquad \text{........ (8)}$$

In Fig. 10B the same slope is shown acted on by a uniform ground acceleration a_H horizontal and perpendicular to the face. The factor of safety now becomes:

$$F = \frac{C\,L + \Sigma\,W_A \cos(\theta_b + \theta_A) \tan\phi}{\Sigma\,W_A \sin(\theta_b + \theta_A)} \qquad \text{........ (9)}$$

and

$$W_A = W/\cos\theta_A$$

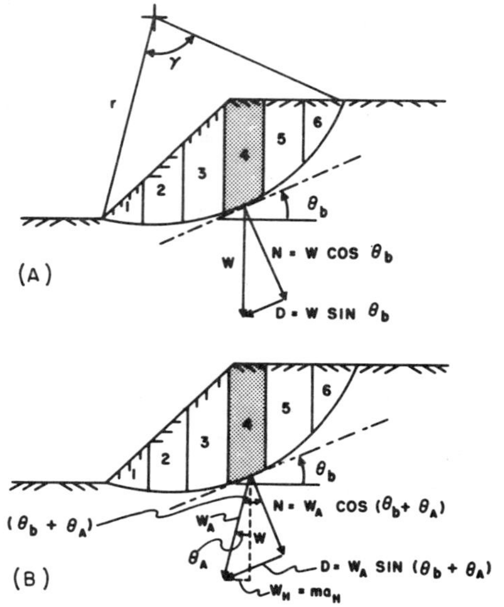

Fig. 10 —Forces acting on a typical slice on an overburden slope (A) in a gravity stress field and (B) in a gravity and blast acceleration field.

It is evident from this equation and from the force diagram in Fig. 10(B) that the driving force is increased markedly while the resisting force is reduced. The net effect is a large reduction in the factor of safety under these conditions.

PRACTICAL IMPLICATIONS OF BLASTING STABILITY ANALYSIS

In the foregoing stability analyses for rock and overburden it was assumed that the potentially unstable mass was acted on by an acceleration of uniform magnitude. In reality the actual wave will consist of positive and negative portions of varying magnitude, each having a finite rise and fall time and approximately sinusoidal in shape. From measurements in rock typically the duration of the main positive acceleration pulse is approximately 0.2 milliseconds. Recognizing that the compression wave velocity of competent rock is about 20,000 ft/sec., the length of the pulse in question is only 4 ft. We can then expect the assumption that a mass is acted on by a uniform positive acceleration to apply only to blocks less than 2 ft. in the dimension perpendicular to the wave. It would be expected that loose blocks this size would be bounced off a wall when acted on by this type of wave and this can be observed in practice. This is not a major stability problem, however, so long as loose and small undercut blocks are not left on pit walls. From the point of view of rock slope stability it is fortunate that these ground pulses are of a high frequency. From Fig. 5 for example, a charge of 10,000 lb. detonated 1000 ft. away would result in an acceleration of 10 g's. It is obvious from the foregoing stability analysis that no slope could withstand a uniform acceleration of this magnitude. Also note from the same figure that for overburden materials the maximum ground acceleration under the same conditions would be 0.1 g's. In overburden the duration of the maximum positive acceleration pulse is typically 10 milliseconds. The compression wave velocity in this material is about 5000 ft/sec. so the length of material being acted on simultaneouly by the positive pulse is roughly 50 ft. From a knowledge of the pulse shape and duration for any particular material, an average uniform acceleration can be assumed and the approach outlined used for stability computations.

In summary then, blast acceleration waves in rock will be of too high a frequency to uniformly accelerate large rock masses causing them to slide. Only pieces less than 2-3 ft. thick will be affected. In overburden, due to the lower frequencies involved, a uniform acceleration of masses up to 50 ft. thick is possible. The effect of this on stability can be analysed as outlined.

Several empirical relationships have been suggested relating particle velocity to damage in rock. Langefors (6) suggest that rock falls in underground openings are likely at particle velocities of 12 in/sec. and rock fracturing at 24 in/sec. Oriard (7) suggests a range of values for rock and soil slopes as follows:

2-4 in/sec. — conservatively safe

25 in/sec. — damage to most rock masses.

In Fig. 11 maximum particle velocity and peak stress at various shot distances are given for small charges of 60% Geogel detonated in blastholes in strong brittle rock. This is taken from a report by Sassa (8). From this we can see that a particle velocity of 25 in./sec. corresponds to a stress level of 2500 psi in this type of material. This is equivalent to the typical tensile strength value of brittle rocks so that the 25 in./sec. criteria seems reasonable in that at this level tensile slabbing could be expected. Above 100 in./sec. the compressive strength of rock would be approached.

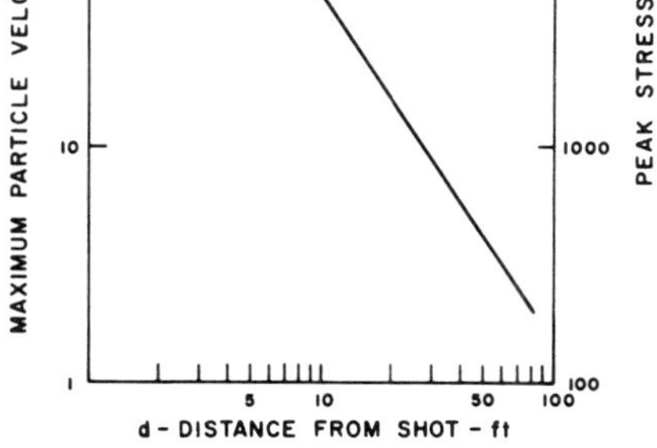

Fig. 11 —Maximum particle velocity and peak stress generated by the detonating of 3400 grams of geocel 60% in a 16.4 cm. hole in strong brittle rock.

The present authors offer the following particle velocity damage criterial for rock masses affected by blasting vibrations.

10 in/sec.	—	no fracturing of intact rock.
10-25 in/sec.	—	minor tensile slabbing will occur.
25-100 in/sec.	—	strong tensile and some radial cracking.
100 in/sec.	—	complete breakup of rock mass will occur.

REFERENCES

1. Edwards, A.T. and Northwood, T.D., "Experimental Studies of the Effects of Blasting on Structures", The Engineer, September, 1960.
2. Duvall, W.I. and Fogelson, D.W., "Review of Criteria for Estimating Damage to Residences from Blasting Vibrations". USBM Report of Investigations, 5968, 1962.
3. Farries, N., Pearson, N., and Anderson, C., Personal Communications with A. Bauer, 1969.
4. Progar, E., "The Effect of Delays on Blasting Vibration Levels", M.Sc. Project, incomplete, Mining Engineering Department, Queen's University, Kingston, November, 1970.
5. Calder, P.N. and Morash, B.J., "Pit Wall Control at Adams Mine". Paper presented at American Mining Congress Metal Mining and Industrial Minerals Convention. Denver, Colorado, Sept. 1970.
6. Langefors, U. and Kihlstrom, B., "The Modern Technique of Rock Blasting". John Wiley and Sons, 1963.
7. Oriard, L.L., "Dynamic Effects on Rock Masses from Blasting Operations", Slope Stability Seminar, University of Nevada, May, 1970.
8. Sassa, K., et al, "The Analysis of Stress for the Prediction of Crater Boundaries", Mines Branch Report R 192, Ottawa, 1966.

Chapter 7

INFLUENCE OF EARTHQUAKES ON STABILITY

by

Robert V. Whitman

Professor of Civil Engineering, Massachusetts Institute of Technology

INTRODUCTION

Ideally, this paper should start by citing examples of slope failures in open pit mines during earthquakes. I do not know of any such examples, although — since my search of the literature has not been exhaustive — such examples may well exist. However, there certainly are many examples of devastating earthquake-induced landslides, and there is no reason why such landslides cannot occur at large open pits. Moreover, many of the actual earthquake-induced landslides have seriously affected mining operations in one way or another. Hence, this paper will treat earthquake-induced landslide phenomena in general, in the hope and expectation that the reader can apply this general knowledge to the specific problems associated with open pit mining.

Ultimately, of course, the aim is to predict the likelihood of the occurrence of earthquake-induced landslides, and to delineate steps which can be taken to minimize the possibility of their occurrence. These aims can be achieved only imperfectly at the present time, for our knowledge concerning earthquake-induced landslides is still in an embryonic stage. Nonetheless, much useful knowledge does exist. In order to understand and apply this knowledge, it is necessary to:

1. Review the nature and extent of actual earthquake-induced landslides;
2. Have a grasp of the basic terminology and concepts of seismology;
3. Study the fundamentals of probability-based seismic risk analysis;
4. Appreciate the way in which earth materials behave during earthquake loadings; and
5. Have a general knowledge of the methods currently available for analyzing the dynamic response of slopes.

Much of this paper is devoted to this necessary background. Information of practical use in connection with various classes of slope stability problems is assembled in the final sections.

LANDSLIDES DURING PAST EARTHQUAKES

Almost every earthquake in mountainous country, strong enough to be reported in the literature, has caused at least some landslides and rockfalls. Unless these slides or falls damage structures or seriously block transportation, they often go unreported. But they do occur. Flores et al (1) describe the effects of such slides and falls during the small Maipo Valley (Chile) earthquakes of 1958 upon mining operations.

With larger earthquakes, such effects become more widespread, and hence are more likely to affect the works of man. The following subsections describe a few of the more important events.

Peru, 1970

The Peru earthquake of May 31, 1970 was by far the most destructive historic earthquake in the Western Hemisphere and ranks high among the world's catastrophic natural disasters (2). As shown in Fig. 1, the epicenter of this earthquake was off the coast. There was widespread damage in the port city of Chimbote (including considerable evidence of liquefaction) and in other towns along the sea coast. However, the greatest devastation occurred in the Callejon de Huaylas, an interior valley between two ranges of the Andes Mountains, at a distance of 80 miles from the epicenter.

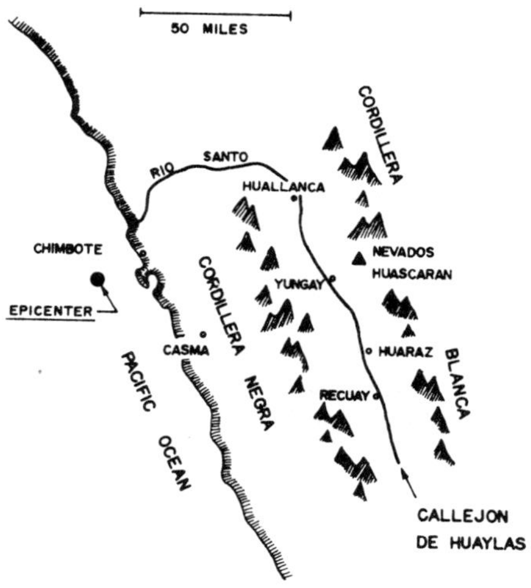

Fig. 1 —Map of zone affected by earthquake of May 31, 1970 in Peru.

The most dramatic devastation was caused by an avalanche from the Nevados Huascaran, the highest peak in the region. This avalanche was triggered by a rockfall or icefall from the very top of this peak; this fall was actually observed by mountaineers climbing on the mountain. This block of ice and/or rock fell over 1000 m. and landed on a glacier. The impact broke up the glacier and set it in motion down the steep slope. As the avalanche roared down the slope at speeds of 100 to 250 mph, rock and soil were added to the mass, until the total volume was about 50 million cu. yd. The avalanche instantaneously buried two towns at the base of the slope. Then the fluid mass turned down the valley of the Rio Santa, and moved at an average speed of 25 mph for at least 80 miles along the valley.

Almost lost in the shadow of this one devastating event were literally thousands of "small" landslides and rockfalls along the steep slopes on both sides of the valley. The existence of these many slides was reported by geologists who flew over the valley soon after the earthquake, and these slides were still obvious during trips along the valley three months after the earthquake. Many of the slides occurred in unconsolidated soils at the bottom of the valley; many others in rocks exposed on the sides of the mountains. These slides contributed to the high death toll in the region, but perhaps their greatest effect was to disrupt access to the valley by road.

Alaska, 1964

The Prince William Sound earthquake of March 27, 1964 was one of the largest historical earthquakes in the Western Hemisphere. It triggered many different types of landslides and rockfalls over a wide area of Alaska. Of these, perhaps the most noteworthy were the slides which took place in stratified sands and clays in the city of Anchorage at a distance of 80 miles from the epicenter (3). Fig. 2, indicates the characteristic nature of these slides. Movement occurred along a horizontal plane. A graben formed where the sliding block broke away from the unaffected soil, and often the block itself broke into many pieces. The largest of these slides — at Turnagain Heights — had an area of 130 acres and moved laterally as much as 600 ft.

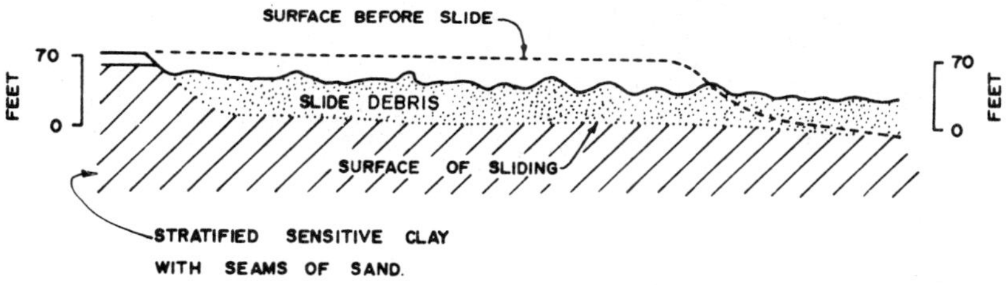

Fig. 2 —Approximate section through slide at Turnagain Heights, Anchorage.

Chile, 1960

The landslides in Anchorage came as a surprise to most engineers, but only because few engineers were aware of the landslides which occurred during the Valdivia earthquake of May 22, 1960. This again was an exceptionally large earthquake, which as usual produced many forms of slope failure (4). The landslide shown in Fig. 3 occurred near Lake Rinihue at a distance of about 80 miles from the fault which produced the earthquake. The soils at Rinihue, like those in Anchorage, were stratified sands and clays of glacial origin. The major effect of this slide, which involved about 30,000,000 cu. yd. of earth, was to dam the river and thereby create a serious problem of flooding when the landslide debris was overtopped. A similar and even larger landslide occurred in this same area during a large earthquake in the early 1500's.

From the experiences in both Alaska and Chile, it is evident that unconsolidated stratified clayey soils are especially susceptible to landslides during earthquakes.

Chile, 1965

The La Liqua earthquake of March 28, 1965 caused numerous landslides and rockfalls, and in addition caused the failure of a number of tailings dams containing wastes from mining operations (5). Fig. 4 shows one of these tailings dams following the earthquake. The contents of the dam, involving nearly 2,000,000 cu. yd. of soil, turned into a viscous liquid, destroyed a town adjacent to the dam and buried more than 200 people, then flowed down a

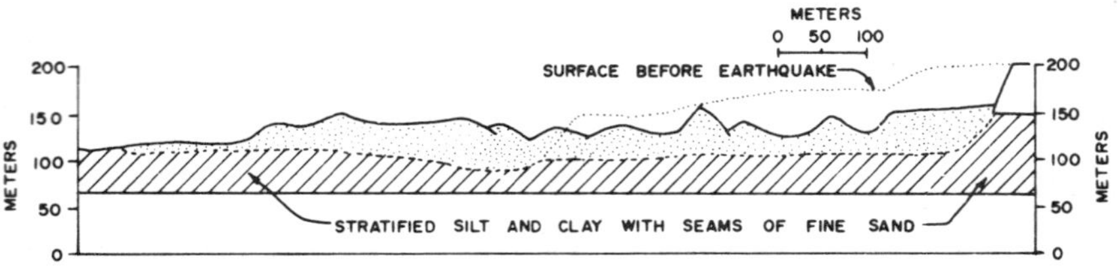

Fig. 3 —Approximate section through slide at Lake Rinihue, Chile.

Fig. 4 —Tailings dam at El Cobre Mine, Chile, after 1965 earthquake.

valley traveling 7 miles in a few minutes. This particular dam was located about 25 miles from the epicenter. In all, eleven tailings dams were totally or partially destroyed during this earthquake.

Montana, 1959

An example of a large landslide in rock occurred in Madison Canyon during the West Yellowstone earthquake of August 17, 1959 (6). Approximately 43,000,000 cu. yd. of rock (schist behind a dolomite dike) broke from the mountainside and created a dam 400 ft. high. This slide buried several vacationers, but fortunately occurred at a location where there were no important man-made structures. The slide took place in the epicentral region, within a mile of the fault along which the earthquake occurred.

CONCEPTS AND TERMINOLOGY OF SEISMOLOGY

This section reviews and illustrates several of the more important concepts and terms.

For a fuller discussion, the reader is referred to **Earthquake Engineering,** published in 1970 by Prentice-Hall and edited by Robert L. Wiegel.

Magnitude

Magnitude is a measure of the size of the earthquake. Theoretically, it would probably make most sense to describe the size of an earthquake in terms of the energy which that earthquake releases. However, there is no way to make a direct measurement of released energy. Hence, an arbitrary definition, suggested a number of years ago by the seismologist Richter, is universally used today. Magnitude on the Richter scale is defined in terms of the motion which would be measured by a standard type of seismograph located one hundred kilometers from the epicenter of the earthquake.

Table 1 gives typical magnitudes of earthquakes which are of particular interest.

Table 1

MAGNITUDES OF SOME IMPORTANT EARTHQUAKES

Location of Earthquake	Year	Magnitude
Very large earthquakes		
San Francisco, Calfornia	1906	8.3
Valdivia, Chile	1960	8.5
Prince William Sound, Alaska	1964	8.4
Earthquakes giving important strong motion records		
El Centro, California	1940	6.7
Parkfield, California	1966	5.5
Other earthquakes mentioned in this paper		
Peru	1970	7.7
La Ligue, Chile	1965	7.1
West Yellowstone, Montana	1959	7.6
Maipo Valley, Chile	1958	5.5

The first three on the list are of interest because of their size; they had magnitudes in the range of 8 or more. In terms of the magnitudes of earthquakes that have been observed in the past, and using extrapolations into the future, most seismologists expect that the size of earthquakes is limited to a magnitude of 9.

The fourth and fifth earthquakes in Table 1 were not of catastrophic magnitude, but they play a particularly important role in earthquake engineering because of actual ground motion measurements that were made during these earthquakes.

Table 1 also gives the magnitudes of the earthquakes whose effects were described in section 2.

Table 2 lists the magnitudes of important earthquakes in eastern and western Canada since 1900.

A word of caution is necessary with regard to magnitude: it is not a precise quantity. Instruments are never located exactly 100 km. from the epicenter, and thus interpolation and extrapolation of actual records is necessary in order to state the magnitude of an earthquake. Hence, there is uncertainty and room for interpretation. For example, the magnitude of the 1970 Peru earthquake was variously interpreted as being anywhere between 7.0 and 7.8. Nonetheless, Richter's magnitude is a very handy and useful measure of the amount of energy released by a given earthquake.

Table 2

MAGNITUDES OF EARTHQUAKES IN CANADA SINCE 1900

from Milne and Davenport (22)

Location	Date	Magnitude
St. Lawrence Region		
Between Arnprior, Ontario and Quyon, Quebec	1924	6.1
St. Lawrence River near the mouth of the Saguenay	1925	7.0
Grand Banks, Newfoundland	1929	7.2
Timiskaming, Quebec	1935	6.25
Cornwall, Ontario	1944	5.9
West Coast Region		
Near Estevan Point	1918	7.0
Queen Charlotte Sound	1929	7.0
Whitehorse, Yukon	1944	6.5
North Strait of Georgia	1946	7.3
Northwestern tip of Queen Charlotte Islands	1949	8.0

Focus and Epicenter

Focus is that point within the earth where the energy from the earthquake is released, whereas the epicenter is the projection of that point upon the ground surface. After an earthquake occurs, the instrumental records are interpreted so as to locate the epicenter or focus. This again is not a precise determination; often there are disagreements of the order of 20 km. as to the location of the epicenter to an earthquake.

In the major earthquake zones of the world (such as the west coast of North America). it is generally agreed that large earthquakes result from slip along pre-existing faults. There is a relationship between the magnitude of the earthquake and the length of fault along which slip occurs. In the Alaskan earthquake of 1964, fault rupture occurred over a distance of about 400 miles (Fig. 5). Earthquakes with a magnitude of 7 are associated with fault rupture over a length of about 25 miles. Earthquakes with a magnitude of 5.5 generally result from rupture along about 4 miles of fault. However, the Parkfield earthquake of 1966, together with its foreshocks, caused rupture extending for about 20 miles.

In such cases, the term epicenter must be used with care. Generally the fault rupture begins at one point of the fault and then spreads along the fault at a velocity on the order of 2 km/sec. The epicenter may not relate well to the damage caused by an earthquake: distance from the causative fault should be a much more meaningful parameter.

Since fault rupture does not occur instantaneously, it might be expected that the duration of strong shaking is related to the length of the fault rupture and hence to magnitude. During the Alaskan earthquake of 1964, significant ground motions at Anchorage continued for 4 minutes or more. On the other hand, earthquakes of magnitude 5.5 cause significant motions only for a few seconds.

For the less seismically active areas of the world, the causes of earthquakes, and the relationship between earthquakes and faults, is not so well understood. Hence, concepts and relations derived from experience in highly seismic areas must be used with caution.

Fig. 5 —Epicenter and fault zone for Alaska earthquake of 1964.

Intensity

Intensity is the measure of the strength of the earthquake as felt at a particular location. From a given earthquake of given magnitude, many people will report different intensities depending upon whether they were close to or at some distance from the epicenter. Hence, whereas magnitude is a measure of the energy released, intensity is a measure of the severity of the shaking or damage at a particular location.

The most commonly used measure of intensity, the so-called Modified Mercalli scale, evaluates intensity subjectively in terms of human reactions and physical damage. For example, Table 3 gives simplified descriptions for four of the categories within the Modified Mercalli Scale of Intensity. Reading through these descriptions, one can see that human evaluation is required to assign intensity.

Fig. 6 is an example of an isoseismal map, in this case from the Parkfield earthquake. Within the epicentral region the intensity was VII, and with distance from the epicenter the intensity dropped until it finally became imperceptible to people. Such a map again emphasizes that intensity is a local affair, and for any earthquake of a certain magnitude many different intensities are reported.

Based primarily upon experience in California, equations and graphs have been developed giving the intensity as a function of the magnitude of an earthquake and the distance from the epicenter. Such relations are used in the estimation of seismic risk but obviously must be used with great caution. There is strong evidence that the same relationship is not applicable everywhere in the world. For example, in eastern North America the rate of decrease of intensity appears to be much slower than in western North America.

The Modified Mercalli scale had its origins in the latter part of the previous century, at a time when there were no instrumental records that could be used for a quantitative measure of intensity, and was developed as a means whereby people interested in earthquake effects throughout the world could communicate and have some common base of describing to each other what was observed. Since much of the historical record of earthquakes is expressed only

Table 3

ABRIDGED MODIFIED MERCALLI INTENSITY SCALE

VI	Felt by all, many frightened and run outdoors. Some heavy furniture moved; a few instances of fallen plaster or damaged chimneys. Damage slight.
VII	Everybody runs outdoors. Damage **negligible** in buildings of good design and construction; **slight** to moderate in well-built ordinary structures; **considerable** in poorly built or badly designed structures; some chimneys broken. Noticed by persons driving motor cars.
VIII	Damage **slight** in specially designed structures; **considerable** in ordinary substantial buildings with partial collapse; **great** in poorly built structures. Panel walls thrown out of frame structures. Fall of chimneys, factory stacks,columns, monuments, walls. Heavy furniture overturned. Sand and mud ejected in small amounts. Changes in well water. Persons driving motor cars disturbed.
IX	Damage **considerable** in specially designed structures; well designed frame structures thrown out of plumb; **great** in substantial buildings, with partial collapse. Buildings shifted off foundations. Ground cracked conspicuously. Underground pipes broken.

in terms of Modified Mercalli intensity, the engineer must be willing to accept and use this scale. Most of the intensities recorded in the literature are based upon sensations experienced by people and effects of ground shaking upon structures. However, in those cases where landslides have caused loss of life or extensive damage, high intensities generally have been assigned. In these latter cases, it is desirable to know the intensities based only upon damage to structures by ground shaking. For example, the intensity at the El Cobre mine during the 1965 earthquake is generally listed as VIII to IX. However, based only on building damage the intensity would be VII to VIII. Table 4 lists intensities in regions where important landslides have occurred, based only upon the effect of ground shaking upon structures.

Table 4

INTENSITIES IN ZONES WHERE LANDSLIDES OCCURRED

Earthquake	Location	Intensity
West Yellowstone, Montana, 1959	Madison Canyon	VII to VIII
Maipo Valley, Chile, 1958	Maipo Valley	VII to VIII
Prince William Sound, Alaska, 1964	Anchorage	VII to VIII
Valdivia, Chile, 1960	Rinihue	VII
La Ligue, Chile, 1965	El Cobre Mine	VII to VIII

Strong Motion Records

Starting in the 1930's, some actual measurements of strong ground motions became available for the first time.

Fig. 7 shows the record, made during the earthquake in 1940, from the instrument located in El Centro, California. It is a record which occurs frequently in earthquake engineering literature, because, in terms of the total strength of the shaking, this is the strongest earthquake record that had been obtained until very recently. Stronger earthquakes have occurred not only before but since this record was obtained, but there were no instruments in the proper locations to measure those motions. For example, there were no instruments to measure the strong motions in Alaska in 1964, or in Chile in 1960.

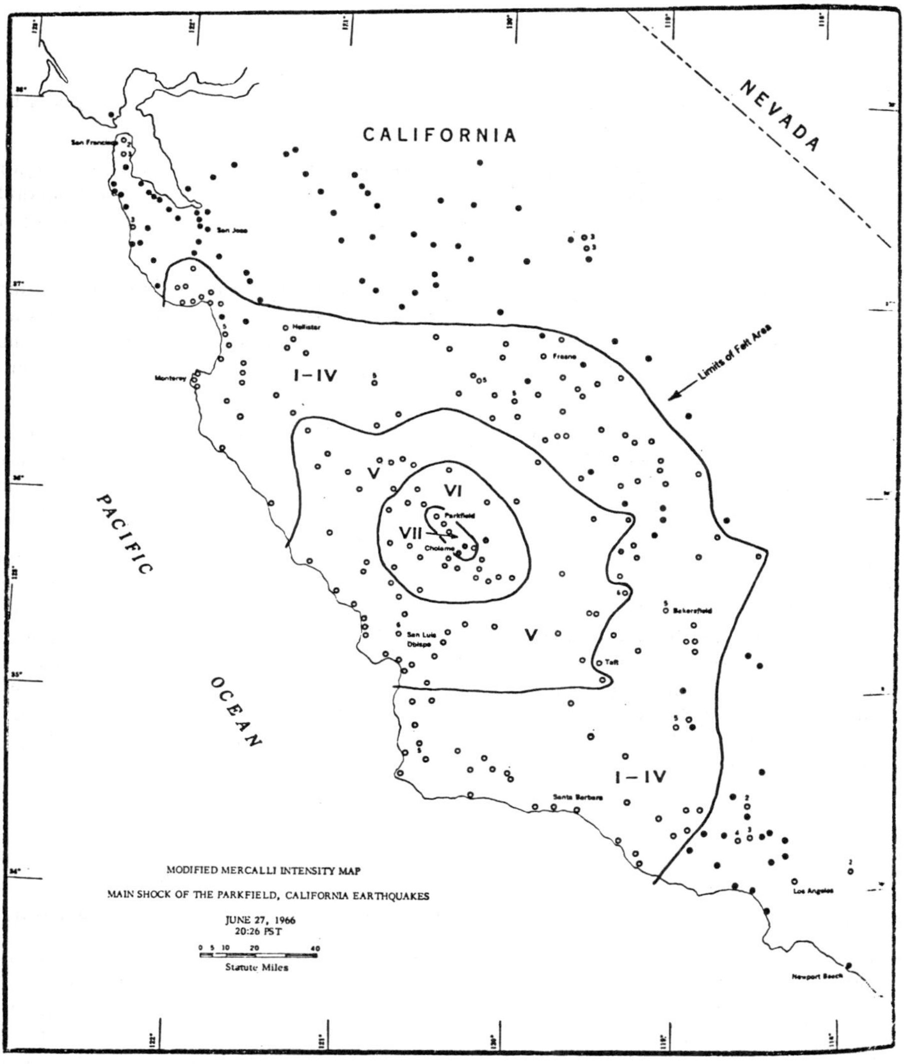

Fig. 6 —Isoseismal map for 1966 Parkfield earthquake (19).

In this record, there are three components of motion: two horizontal components plus the vertical component. The peak horizontal accelerations go up to about 0.33 g. The strong motions in the horizontal directions lasted for some 25 to 30 sec., although the shaking certainly was not of uniform intensity throughout this period. The vertical component also shows, in the very initial stages, some large accelerations in excess of 0.2 g, although looking at the record as a whole the intensity of shaking in the vertical direction is not as strong as that in the horizontal directions.

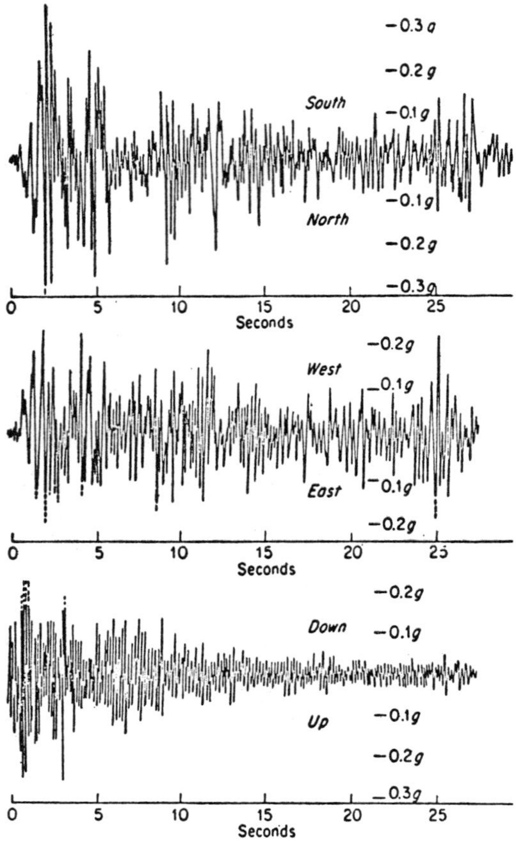

Fig. 7 —Record from El Centro accelerograph during 1940 earthquake (20).

Much is learned by integrating an acceleration record to obtain the velocity and the displacement of the ground as a function of time. In Fig. 8, this has been done for the north-south component of the 1940 El Centro earthquake record. The peak velocity is on the order of 15 in./sec, the peak displacement is a little over 8 in. It is important to recognize the difference in the character among acceleration, velocity, and displacement records. There are many high-frequency components in the acceleration records, and it is these high-frequency components which carry the large accelerations. There also are many low-frequency components as well; it is these low-frequency components, when integrated with respect to time, that account for the observed displacement.

Fig. 9 shows the recording from a different earthquake: the Parkfield earthquake which occurred between San Francisco and Los Angeles in 1966. This was a very short duration earthquake, measured almost literally right at the epicenter of this magnitude 5.5 earthquake. The peak measured acceleration in the horizontal direction was about 0.5 g. When integrated, this record produced large velocities (on the order of 30 in/sec) and large displacements (as large as 8 to 10 in).

The El Centro and Parkfield records represent two extreme types of records: one with many pulses of motion extending over many seconds, and the other with only a single pulse lasting just a few seconds. Many other records have been obtained which fall between these extremes, and of course there are many records involving less intense motions. Within the last several years, there has occurred at least one other earthquake which produced accelerations greater than 0.5 g: the earthquake at Konya Dam in India. Various engineers who have studied the measured motions suggest that the peak acceleration was between 0.5 and 0.65 g.

As actual strong motion records became available, various attempts have been made to

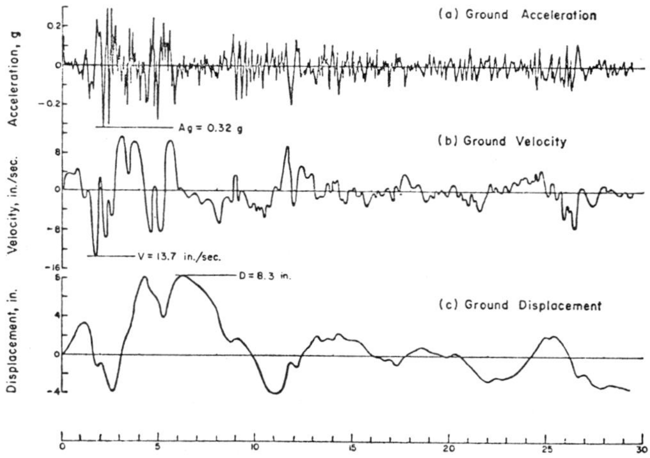

Fig. 8 —Motions computed from N-S component of 1940 El Centro record (15).

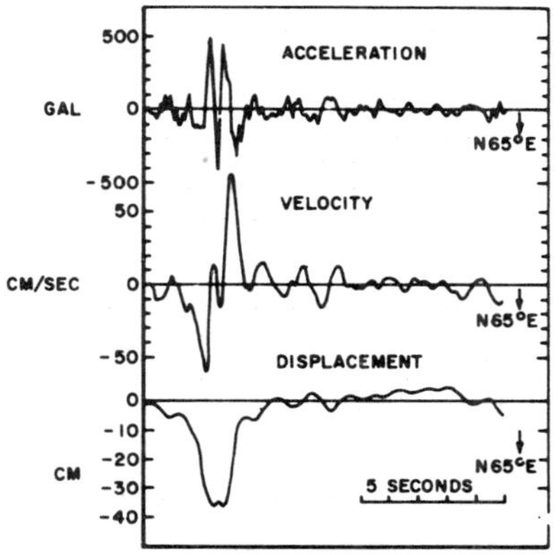

Fig. 9 —Accelerographs from Parkfield earthquake (21).

correlate peak acceleration (and peak velocity) with the magnitude of the earthquake and the distance from the epicenter. Esteva (7) presents a recent correlation. Equations describing these correlations play an important role in the analysis of seismic risk. However, because of various geological factors, there is considerable scatter in the data upon which these correlations are based. The existence of this scatter must be recognized when using these correlations.

Damage Correlations

Most engineers would prefer to have the intensity of ground motion described by a number, such as peak acceleration, rather than by the subjective Modified Mercalli (MM) scale. Hence, since ground motion measurements became available, there have been numerous attempts to correlate ground motion with MM intensity.

Fig. 10 shows two oft-used correlations between peak acceleration and MM intensity. The figure also shows that there is a huge scatter in the data. Two examples serve to illustrate the difficulty.

1. The MM intensity was VIII at the site where the 1940 El Centro record (0.33 g) was obtained, whereas the MM intensity was only VII at the site where the 1966 Parkfield record (0.5 g) was recorded.
2. During the 1952 Southern California earthquake, three strong motion records were obtained at three sites where the damage was rated as MM intensity VII. The measured peak accelerations ranged from 0.05 g to 0.17 g.

Clearly, peak acceleration does not correlate well with damage as rated by the MM intensity scale.

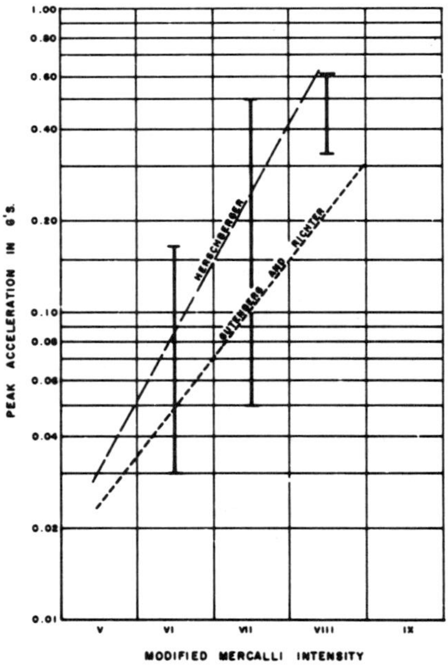

Fig. 10 —Correlation of peak acceleration and modified Mercalli intensity.

Alternatively, it has been suggested by Neumann that intensity should be correlated with peak velocity. However, this correlation proves to be only slightly better than that in Fig. 10.

Indeed, it is clear that no one single feature such as peak acceleration or peak velocity can correlate well with damage. At least one other feature must be considered: the duration of time over which the shaking occurs and the number of pulses involved in the motion. The Parkfield earthquake illustrates this point. This earthquake, with its single sharp pulse, did cause damage in the form of cracks through abutments of bridges. However, since no collapses occurred and the costs of repair were minor, damage was rated as slight. However, if this earthquake had lasted only slightly longer, with several cycles of loading, the cracked abutments would have distorted severely and even possibly collapsed, and the MM intensity for this earthquake would have been at least VIII.

Correlations such as those in Fig. 10 are often used in seismic risk analysis, and must be regarded with considerable skepticism. As more strong motion records are obtained, it can be expected that improved correlations will be developed, although more than one single feature of the ground motion will be involved. Arias(8) has suggested a promising new form of correlation, but as yet this approach has not been tested in practice.

SEISMIC RISK ANALYSIS

It is the writer's belief that any rational assessment of earthquake threat, or any rational design against earthquake effects, must be based upon probabilistic analysis. This belief is neither widely recognized nor widely accepted, as is illustrated by the following observations.

1. Most building codes are not based directly upon probabilistic considerations. These codes are based upon sound judgment by engineers who may have probabilistic considerations in the back of their minds. However, most codes involve unassessed risks and unknown conservatisms.

2. Design of nuclear power plants is still based upon the concept of a maximum credible earthquake — a rather vague concept which defies further definition. The designs for these plants are safe and sound, but involve conservatism which is unknown and which varies from plant to plant.

Most of the literature of earthquake engineering comes from engineers who design buildings, and such engineers traditionally are loath to admit that there is any risk of failure. Fortunately this situation is changing.

Seismic risk analysis based upon probabilistic considerations basically involves two steps:

1. Establishing the probability that various levels of earthquake ground shaking will occur.

2. Choosing an acceptable level of probability as a basis for design.

The second of these steps is probably the more difficult step, involving assessment of: the probability that a failure will occur, given that an earthquake of a given size occurs; the penalties associated with failure, and the costs necessary to prevent a failure. However, this section will concentrate upon the first of these steps.

The first step usually is accomplished by constructing a graph of strength of ground shaking vs. return period; an example is shown in Fig. 11. An earthquake with a return period of 50 years has a probability of 100% of occurring once in a 50 year interval.

The strength of ground shaking may be expressed in terms of MM intensity or in terms of some quantitative factor such as peak acceleration. There are advantages to both representations. Engineers usually prefer to have a quantitative factor, but generally the raw data is in the form of intensity.

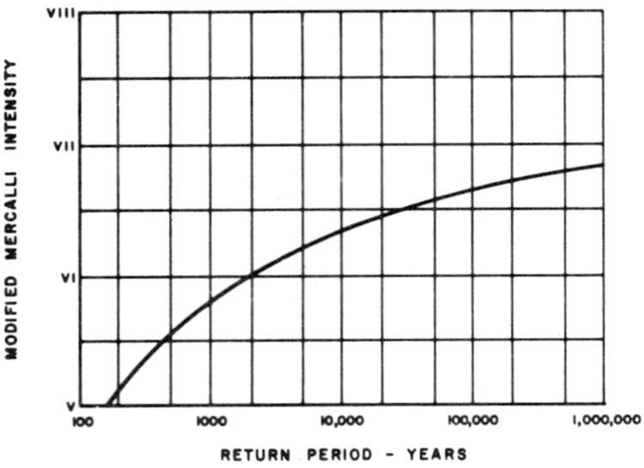

Fig. 11 —Example of return period graph.

Several different procedures may be used to construct the return period graphs; these procedures are outlined in Table 5. In the rare case where sufficient observations of actual earthquakes are available at the site, a level 1 approach generally is best. Level 2 and level 3 approaches may be used to advantage when actual observations at a site are of short duration or are incomplete. Approach 3C (7) is the most sophisticated and best approach.

The results of such analyses are illustrated in Figs. 12 through 15. The curves for eastern Canada were prepared using approach 2A. The raw data are in the form of MM intensities and the conversion to acceleration was made as a final step. The interrelationship between intensity and acceleration is as follows:

Acceleration	**MM intensity**
0.03 g	VI
0.06 g	VII
0.10 g	VII to VIII
0.20 g	VIII to IX

The curves for western Canada were prepared using approach 2B. Thus the dispersion in the attenuation laws was not taken into account, and the figures underestimate the probabilities of occurrence. Also, the calculations were made directly in terms of acceleration rather than in terms of MM intensities. The reader must make his own choice of correlation from the information in Fig. 10.

BEHAVIOUR OF SOIL AND ROCK DURING EARTHQUAKES

At some risk of oversimplification, slope stability problems during earthquakes may be divided into two categories.

1. Those in which the pore pressures within the soil or rock increase steadily with each cycle of straining during the earthquake ground motion.

2. Those in which there is little or no increase in pore pressure.

The first of these categories gives rise to the most spectacular slope failures during earthquakes.

The build-up of pore pressure during repeated cycles of loading has been demonstrated

Table 5

APPROACHES FOR COMPUTATION OF RETURN PERIODS

Level	**Method**	**In terms of MM INTENSITIES**		
1	Assemble observations made directly at site	1A	1B	
			Using average attenuation law	**Considering scatter in attenuation law**
2	Use known epicenters; use attenuation laws relating magnitude, intensity and distance to compute intensities at site	2A	2B	2C
3	Determine probability of various magnitudes of earthquake occurring at various locations along fault or within zone; use attenuation laws to compute intensities at site	3A	3B	3C

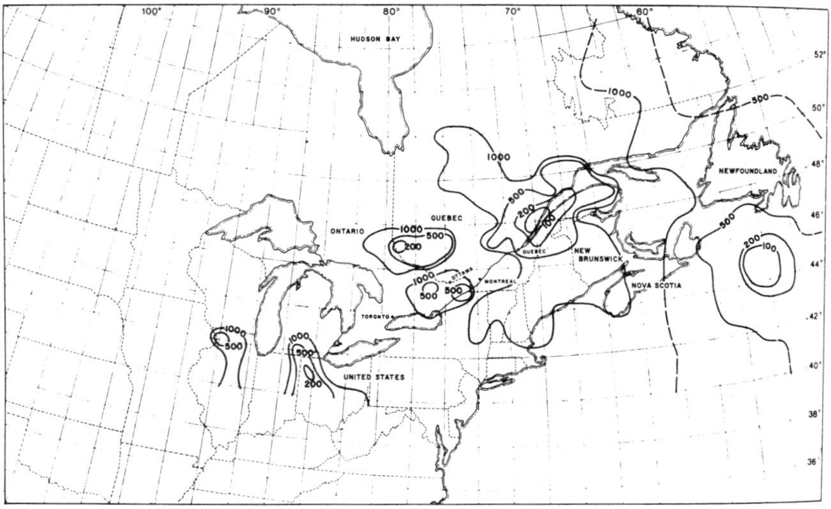

Fig. 12 —Contour map of return period in years for acceleration of 0.1 g. in Eastern Canada (22).

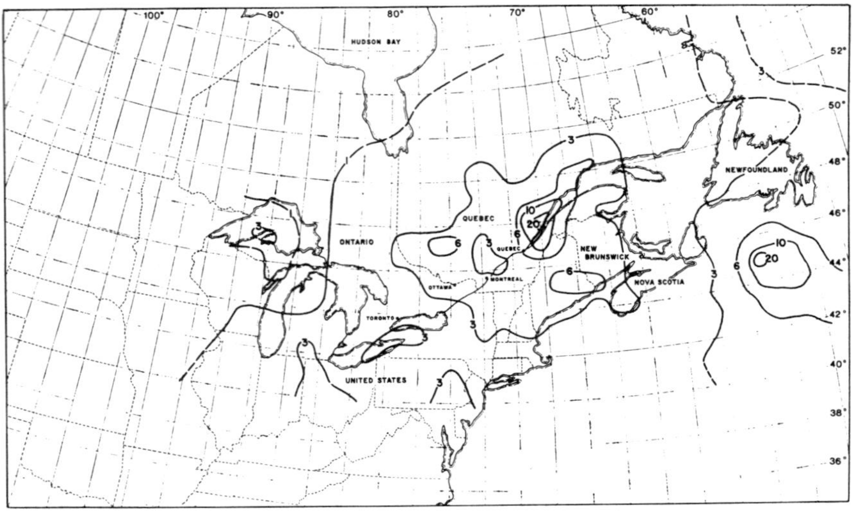

Fig. 13 —Contour map of accelerations as a per cent of g with a 100 year return period for Eastern Canada (22).

by laboratory triaxial tests on loose sands. Fig. 16 shows a typical set of results. As the sample is subjected to a cycle of increasing and then decreasing axial stress, a net increase in pore pressure occurs. The pore pressure continues to increase until it approaches the total confining stress applied to the sample. At this point, since the effective stress between the sand particles has dropped to a low level, the shear resistance of the sand becomes very small. The sand thus no longer can resist the applied stress, and very large strains occur. When the sand reaches this condition, it often is said that the sand has liquefied, although there is dispute as to whether this is appropriate terminology.

The effects of such pore pressure build-up and loss of strength can be dramatic(9). During the 1964 earthquake in Niigata, Japan, the sand had its bearing capacity destroyed and buildings tipped and sank into the soil. The tailings dams in Chile experienced liquefaction during the 1965 earthquake. Where strength loss occurs within slopes, slope failure may result. Loss of strength either within seams of sand or within sensitive clays accounted for the slope failures in Anchorage (1964) and Rinihue (1960).

Thus prediction as to whether or not such a loss of strength can occur is often vital to the rational assessment of slope stability during an earthquake. Such a prediction involves two steps(10):

1. Prediction of the magnitude and number of cycles of dynamic stress which will occur in the soil during an earthquake.

2. Tests to determine whether or not this stress (both magnitude and number of cycles are important) will cause large buildup of pore pressures and loss of strength.

The first step utilizes mathematical analysis: while there are uncertainties in this analysis, reasonable estimates can be made for the dynamic stresses associated with a given ground motion. However, there is at present great uncertainty and controversy regarding test procedures which have been developed for the second step.

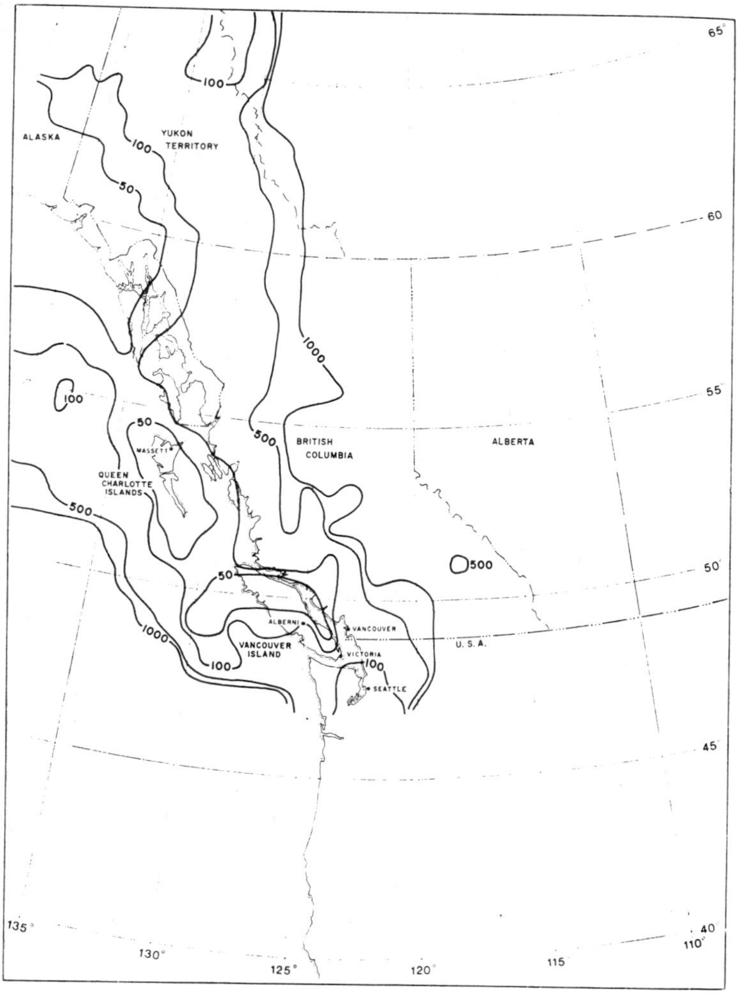

Fig. 14 —Contour map of return period in years for acceleration of 0.1 g in Western Canada (22).

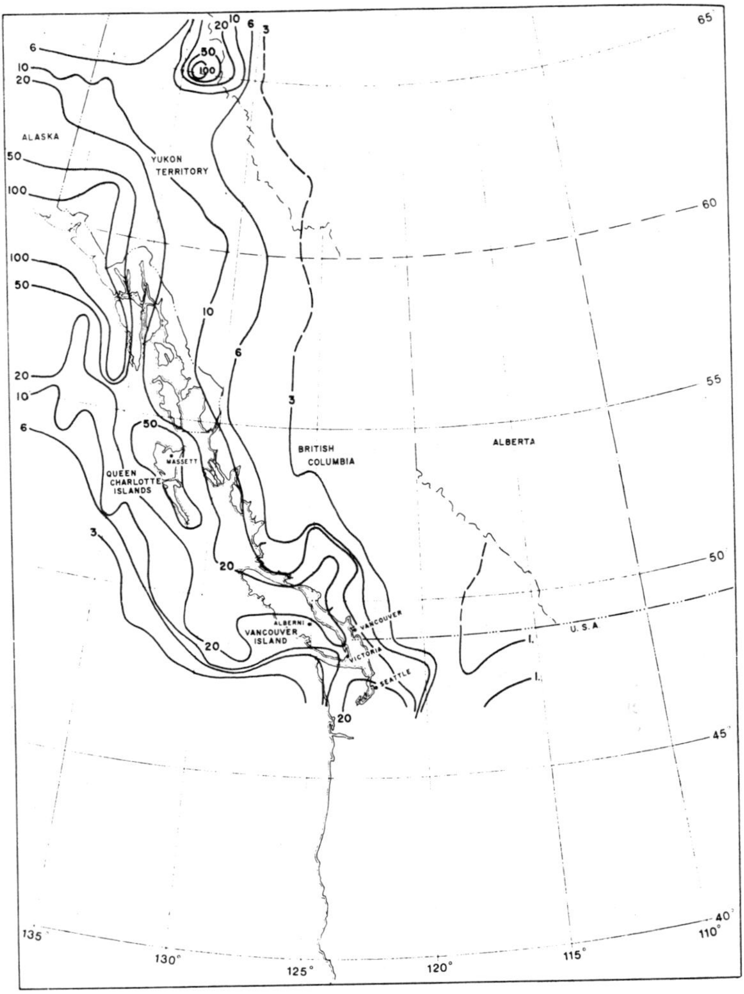

Fig. 15 —Contour map of accelerations as a per cent of g with a 100 year return period for Western Canada (2)).

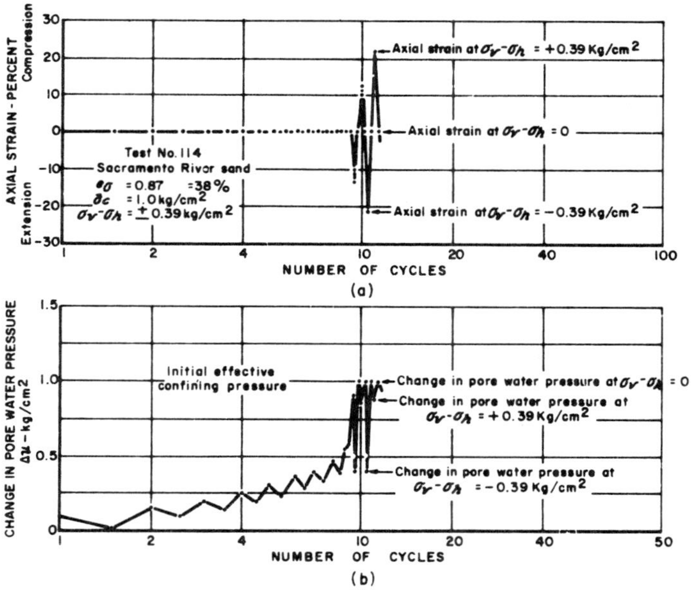

Fig. 16 —Results from repeated load triaxial test on fine sand (11).

The original suggestions regarding test procedures were made by Seed and Lee(11) and Peacock and Seed(12). Unfortunately, the combination of stress and number of cycles required to cause liquefaction of a given sand has proved to be quite sensitive to test conditions; the recent paper by Finn et al(13) is the latest in a series of papers dealing with the effect of test conditions. Moreover, the validity of all such laboratory tests has been questioned on the basis that these tests fail to simulate field conditions(14).

Because of these difficulties it is not at the present time possible to make an accurate determination of the behavior of soils (cohesionless soils and sensitive clays) which are susceptible to a liquefaction-like phenomena. About the best that can be done is to determine a range for the stress necessary to cause liquefaction in a given number of cycles. The upper limit of this range may well be twice the lower limit.

Fortunately, not all soils are susceptible to such phenomena. For many soils, the resistance to shear is largely unaffected by repeated cycles of loading. Despite the difficulties with the laboratory tests, they are very useful for determining whether or not a given soil is susceptible to liquefaction.

ANALYSIS OF DYNAMIC RESPONSE OF SLOPES

If a liquefaction occurs within a slope, the movements of the slope can be extremely large. The amount for a given earthquake is affected by many details of the circumstances, and hence is difficult to predict. Hence, a prediction that liquefaction will occur amounts to a prediction that catastrophic motions will occur.

If the soil (or rock) within a slope does not experience a loss of strength, it is possible to estimate the magnitude of the motions which the slope will experience if it is overstressed during an earthquake. Newmark(15) has developed a simple analysis for this purpose.

The key features of this analysis may be examined by considering the problem of a block resting on an inclined plane (Fig. 17). If the block is to accelerate in a direction parallel to the slope, the shear force between the block and the slope must differ from T, the shear force required for static equilibrium. Since the shear force is limited (it cannot exceed N tan θ , where N is the normal force and θ the friction angle), the acceleration the block can experience is limited:

maximum upslope acceleration:

$$\frac{W}{g} A^1 g = W \cos\theta \tan\phi - W \sin\theta \qquad \text{. . . . . (1)}$$

or

$$A^1 = \cos\theta \tan\phi - \sin\theta \qquad \text{. . . . . (2)}$$

maximum downslope acceleration:

$$A^1 = \cos\theta \tan\phi + \sin\theta \qquad \text{. . . . . (3)}$$

If the maximum acceleration coefficient A of the slope is less than A^1, then the block and the slope will move together without relative displacement. However, if $A^1 < A$, then relative displacement will develop. The block cannot keep up with the slope as the slope accelerates uphill, and hence relative downhill displacement occurs. During downhill acceleration the block and slope are able to move together since A^1 is greater in this case. The total downhill move-

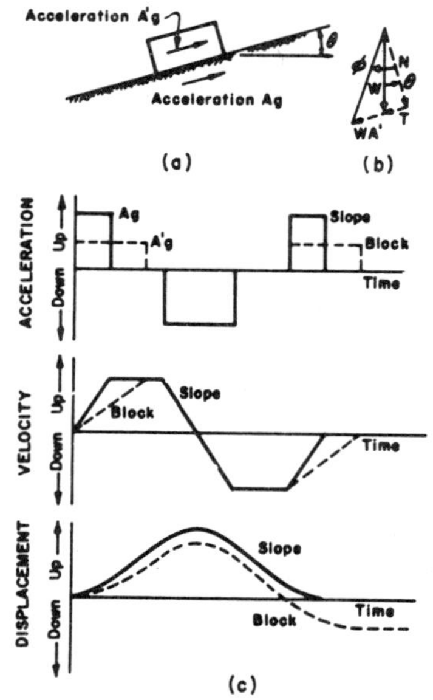

Fig. 17 —Motion of block on inclined plane during transient loading.

ment as a result of an earthquake motion with a series of acceleration pulses is obtained by summing the movements during the individual pulses. The method may be extended to apply to slope failure conditions, including rotational type failures. Newmark's paper contains charts which may be used to obtain quick estimates of the total movement during a given earthquake. The general correctness of the theory has been illustrated in model tests by Goodman and Seed(16).

An alternate method for analyzing the dynamic response of slopes has been proposed by Seed(17). This method, which is especially useful with earth dams, is applicable when the expected movements are small.

SLOPES IN SOIL WHERE LIQUEFACTION IS NOT A PROBLEM

Now that the necessary background information has been reviewed, some practical recommendations are possible concerning various categories of stability problems. The first, and simplest, category involves soils which will not lose strength during an earthquake because of liquefaction-type behavior.

In such cases, theory and experience have shown that large slope movements occur only when the static factor of safety is less than about 1.3. When the static factor of safety is small — 1.05 or 1.1 — large movements can and do occur if an earthquake is strong and of long duration. It is possible to estimate the magnitude of these movements using the theory described previously. By combining calculations using this theory and a probabalistic assessment of expected ground motions, curves giving expected movement as a function of return period can be constructed. Then rational decisions can be made concerning the possible future costs associated with earthquake-induced failures and the present costs of possible steps to minimize the future risks.

Two specific types of practical situations fall within this category.

Dry Slag Piles

Piles of nearly dry cohesionless materials are often made by simple dumping, and the slopes of such piles usually are at the angle of repose for the material. That is to say, the static factor of safety for these piles is unity, or only slightly greater than unity. Hence such piles may experience large movements during earthquakes. During earthquakes of MM intensity VII, such piles have spread to twice their original diameter while settling to half of their original height. The potential future costs for such earthquake-induced spreading take the form of damage to equipment and structures located at the base of dumped piles, and loss of life to personnel on the equipment or in the buildings. Since it probably is not practical to maintain these piles at slopes less than the angle of repose, steps to reduce the earthquake hazard take the form of moving equipment and structures to a greater distance from the pile.

It must be emphasized that slag piles which contain water near their bottom are susceptible to liquefaction and hence are an entirely different, and much more serious, problem. Suggestions made in the section on tailings dams will apply to such cases.

Slides along existing failure surfaces

In nature one often finds slopes which have already failed and moved to some extent, but which are now stable with a safety factor slightly greater than unity. Such slopes must be expected to move during strong earthquakes. Movements can become important during earthquakes of intensity VII, and can become large when the intensity reaches VIII, especially for earthquakes of long duration. The possible movement of such slopes can be evaluated as a function of return period.

Where such slopes occur along reservoirs, the potential serious consequence of a large movement is obvious, and definite steps may be justified to make a slope more stable (by better drainage of groundwater) or to provide more freeboard in the dam which creates the reservoir.

A much more common consequence of such slope movements is disruption of transportation routes. This type of failure usually represents a risk which must simply be accepted, although provision for alternate access routes may be desirable in cases where the consequences of disruption are unusually severe.

TAILINGS DAMS

The fines in the interior portions of typical tailings dams are very susceptible to liquefaction during earthquakes. The most practical approach to minimizing the danger inherent in tailings dams is to have an exterior shell which will contain the liquefied interior portion.

Unfortunately, the exterior portions of many tailings dams consist of saturated, loosely-deposited sands, and hence are themselves subject to liquefaction type failures. Based upon the experience in Chile, significant probability of failure of such tailings dams exists when the MM intensity reaches VII, and the probability of failure exceeds 50% when the MM intensity reaches VIII. Where such earthquake threats exist, serious consideration should be given to strengthening existing tailings dams and new dams should be constructed with exterior shells that will resist failure.

Another paper to this conference by Dr. L. Casagrande discusses practical methods for improving the performance of tailings dams. The most obvious steps are to compact the soil in the outer shell and to provide drainage to lower the phreatic level in the outer shell. It also is important to avoid sharp corners where tension cracks can develop in the exterior shell during earthquake shaking.

Detailed dynamic analysis and strength testing, following the procedures suggested by Seed (17), can be of valuable aid in the evaluation of the resistance of the exterior shell. However, the limitations discussed earlier must be kept in mind.

SLOPES IN STRATIFIED SANDS AND CLAYS

This category represents an important but (for the present at least) frustrating type of problem. Sands, silts and sensitive clays of relatively recent geological origins, and deposits with interstratifications of these soil types, are very common in mining areas. As discussed previously, there have been notable failures of slopes in such soils during earthquakes. Considerable research has been devoted to the testing and behavior of these soils, and to the analysis of slopes composed of these soils. Yet, an accurate assessment of the stability of these slopes is not possible at this time.

The historical record indicates that failures of slopes in such soils begin to occur when the MM intensity reaches VII. With earthquakes of this intensity but short duration, failures will be infrequent; peak accelerations must reach 0.2 g or 0.3 g to cause these failures. On the other hand, massive failures must be expected during ground motions of MM intensity VII but long duration. Where the duration is very long, peak accelerations of only 0.1 g may cause failure. The experience in both Alaska and Chile shows that massive failures occur within 100 miles from the causative fault during an earthquake of magnitude 8.

A rational decision concerning such a slope can be made only on the basis of a very thorough geotechnical investigation, involving numerous borings and many laboratory tests. This investigation must include a detailed study of the record of stability of slopes in the immediate region. As noted, it may not be possible to achieve high accuracy in the predictions, but a thorough investigation can nonetheless provide the basis for a sound decision. The expense of the investigation is justified whenever a failure would cause loss of life or serious economic loss.

It should also be noted that rather small earthquakes may contribute to the deterioration of a slope which already is on the verge of failure. Brzezinski (18) has provided a recent example.

ROCK SLOPES

The state-of-the-art is exceedingly poor with regard to this category of problem. Rockfalls begin to occur when the MM intensity reaches VII, and the frequency of rockfalls and rockslides increases with increasing intensity. Such falls and slides undoubtedly only occur when the static safety already is close to unity, and the ability to predict failure during an earthquake cannot possibly be better than the ability to predict static safety. If the consequences of a failure are serious, if the equilibrium under static conditions is precarious and if earthquakes of intensity VII are frequent, then it can be said that there is a serious earthquake problem.

SUMMARY

The main thoughts expressed in this paper may be summarized as follows:

1. The threat posed by earthquakes should be approached by the following steps:
   a. Determining the expected intensity of ground motions on a probabalistic basis.
   b. Assessing the likelihood of failure for different levels of ground motion.
   c. Analyzing the potential future costs associated with possible earthquakes.
   d. Ascertaining the cost of present actions to forestall the possible future costs.
2. Step b is the most difficult step. Crude estimates may be used for preliminary analysis. More sophisticated analysis will be justified in those cases where the potential future costs and the cost of present remedial action are both high.
3. Except in cases where liquefaction-type behavior loss of strength can occur, slope failure during earthquakes occurs only when the static safety factor is low. If the static safety factor can be evaluated, then a rational assessment of the likelihood and extent of failure during an earthquake is possible.
4. Where liquefaction can occur, failure during an earthquake is possible even when the static safety factor is high. Accurate prediction of liquefaction is not possible at present, but careful exploration, testing and analysis can provide a basis for rational evaluation of the risk.

REFERENCES

1. Flores, R., Arias, S., Jenschke, V., and Rosenberg, L.A., "Engineering Aspects of the Earthquakes in the Maipo Valley," Proc. 2nd World Conference on Earthquake Engineering, Vol. I, Tokyo, 1960, pp. 409-433.
2. Ericksen, G.E., Plafker, G. and Concha, J.F., "Preliminary Report on the Geologic Events Associated with the May 31, 1970, Peru Earthquake," U.S. Geological Survey Circular 639, 1970.
3. Wilson, S.D., "Landslides in the City of Anchorage," The Prince William Sound, Alaska, Earthquake of 1964 and Aftershocks, Vol. II, Part A, U.S. Government Printing Office, Washington, D.C., 1967, pp. 253-297.
4. Duke, C.M. and Leeds, D.J., "Response of Soils, Foundations and Earth Structures to the Chilean Earthquakes of 1960," Bull. Seismological Society of America, Vol. 63, No. 2, 1963.

5. Dobrey, R. and Alvarez, L., "Seismic Failure of Chilean Tailings Dams," Proc. Journal Soil Mechanics and Foundations Division, ASCE, Vol. 93, No. SM6, 1967, pp. 237-260.

6. Barney, K.R., "Madison Canyon Slide," Civil Engineering, Vol. 30, August, 1960, pp. 72-75.

7. Esteva, L., "Seismic Risk and Seismic Design Inputs," Seismic Design for Nuclear Power Plants, R.J. Hansen, ed. M.I.T. Press, 1970, pp. 142-182.

8. Arias, A., "A Measure of Earthquake Intensity," Seismic Design for Nuclear Power Plants, R.J. Hansen, ed. M.I.T. Press, 1970, pp. 438-483.

9. Seed, H.B., "Landslides During Earthquakes Due to Liquefaction," Proc. Journal Soil Mechanics and Foundations Division, ASCE, Vol. 94, No. SM5, 1968. pp. 1055-1122.

10. Seed, H.B. and Wilson, S.D., "The Turnagain Heights Landslide, Anchorage, Alaska," Proc. Journal Soil Mechanics and Foundations Division, ASCE, Vol. 93, No. SM4, 1967, pp. 325-353.

11. Seed, H.B. and Lee, K.L., "Liquefaction of Saturated Sands During Cyclic Loading," Proc. Journal Soil Mechanics and Foundations Division, ASCE, Vol. 92, No. SM6, 1966, pp. 105-134.

12. Peacock, W.H. and Seed, H.B., "Sand Liquefaction Under Cyclic Loading Simple Shear Conditions," Proc. Journal Soil Mechanics and Foundations Division, ASCE, Vol. 94, No. SM3, 1968, pp. 689-708.

13. Finn, W.D.L., Bransby, P.L. and Pickering, D.J., "Effect of Strain History on Liquefaction of Sand," Proc. Journal Soil Mechanics and Foundations Division, ASCE, Vol. 96, No. SM6, 1970, pp. 1917-1934.

14. Castro, G., "Liquefaction of Sands," Ph.D. thesis, Division of Engineering and Applied Physics, Harvard University, 1968.

15. Newmark, N.M., "Effects of Earthquakes on Dams and Embankments," Geotechnique, Vol. 15, 1965, pp. 139-160.

16. Goodman, R.E. and Seed, H.B., "Earthquake-Induced Displacements in Sand Embankments," Proc. Journal Soil Mechanics and Foundations Division, ASCE, Vol. 92, No. SM2, 1966, pp. 125-146.

17. Seed, H.B., "Earth Slope Stability During Earthquakes," Earthquake Engineering, R.L. Wiegel, ed., Prentice-Hall, 1970, pp. 383-401.

18. Brzezinski, L.S., "A Review of the 1924 Kenogami Landslide," Canadian Geotechnical Journal (in publication), 1971.

19. Cloud, W.K., "Intensity Map and Structural Damage, Parkfield, California, Earthquake of June, 27, 1966," Bull, Seismological Society America, Vol. 57, No. 6, 1967, pp. 1161-1178.

20. Anderson, A.W. et al., "Lateral Forces of Earthquakes and Wind," Trans., ASCE, Vol. 117, 1952.

21. Housner, G.W. and Trifunac, M.D., "Analysis of Accelerograms — Parkfield Earthquake," Bull. Seismological Society of America, Vol. 57, No. 6, 1967, pp. 1193-1220.

22. Milne, W.G. and Davenport, A.G., "Distribution of Earthquake Risk in Canada," Bull. Seismological Society of America, Vol. 59, No. 2, 1969, pp. 729-754.

Chapter 8

METHODS OF ANALYSIS OF STABILITY OF ROCK SLOPES

by

David L. Pentz

Senior Engineer, Golder, Brawner and Associates, Vancouver

ABSTRACT

A critical summary is given of the analytical procedures available to the engineer for the analysis of slope stability.

The emphasis is on the various types of behaviour that have been measured in different rock types.

A method of assessing the possibility of failure of simple three dimensional wedges is provided in an attempt to demonstrate how structural features can be included in a meaningful manner.

An outline for theoretical design is provided that stresses the importance of carrying out a 'Back Analysis' of large and small instrumented failures so that realistic parameters of the rock mass forming the pit slopes can be obtained.

INTRODUCTION

The slope geometry of an open pit mine affects the way in which the ore reserves are removed. It becomes critical if a reasonable estimate has to be made of the anticipated size and/or life of a mine.

Slopes in open pit mines fall into three categories (Fig. 1):

(i) 'Total' pit geometry

(ii) Operating pit geometry

(iii) Bench geometry

The 'Total' pit geometry is defined by an economic comparison between the cost of removing and processing a greater percentage of the ore bearing material in the pit area with the benefits provided from selling the additional ore. Thus, the 'Total' pit is subject to the price of the minerals mined and therefore, for feasibility studies, it is largely a fictitious concept, except where the expected life of a mine is very short.

The operating pit geometry has, until recently, been defined by an analysis of operating costs. The overall operating slopes in most large pits vary from 20° to 30°. In some cases, an economic examination has shown that an increase in the operating slopes can improve the cash

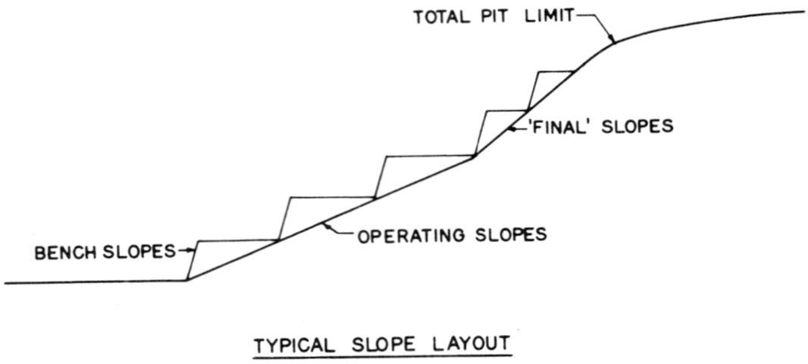

Fig. 1 —Typical slope geometrics of open pits.

flow since less waste is removed over a given period of time. The advantages of steeper slopes must be balanced against the probable reduction in operating efficiency and resultant increase in operating costs due to the more sophisticated drilling, blasting, and general mining techniques sometimes required by steeper slopes. In fact, steeper final slopes, in many cases, may require the purchase of different equipment.

The bench geometry is also partially controlled by operating conditions. The height of the bench which normally ranges from 25 to 100 ft., and the bench width define the operating slope angle. The stability of benches, particularly of final wall benches, is sometimes critical to a pit since excessive failure of benches can seriously inhibit the safety of the operations. This effect is most noticeable in small pits.

This paper has been prepared to outline the methods of assessing slope behaviour. If a 'Stability Analysis' is to be useful, it must satisfy the following criteria:

(i) The real conditions of the rock mass forming the slope must be adequately modelled so that the likelihood of failure for a given slope geometry can be assessed.

(ii) Practical experience must confirm theoretical predictions.

These conditions have been satisfied fairly well for problems involving soils. Such materials are often found on the perimeters of open pits. The success of soil mechanics principles has resulted in similar techniques being applied to the underlying rock. In some instances, modifications of soil mechanics techniques have proved successful, but in general the effects of the geological discontinuities in rock masses have necessitated more complex techniques of analysis.

OUTLINE OF ANALYTICAL PROCEDURES

A satisfactory theoretical analysis must account for all the various parameters that control stability. These parameters may be developed in three stages.

Stage 1

The geology of the materials forming the slope must be sufficiently well delineated so that the lithological boundaries of the different rock types can be determined. This also involves a structural investigation to determine the spatial relationships between the discontinuities. The discontinuities being the faults, shears, joints, bedding planes, foliation, etc. Broadbent and Rippere, (1) Robertson, (2) Rosengren, (3) and Terzaghi (4) have discussed mapping techniques in detail.

Stage II

The imposed loading brought about by self-weight (gravity), water pressure, residual tectonic stress, seismic loads due to blasting or earthquakes must be determined.

Stage III

The available strengths, or alternatively, the allowable displacements or deformations of the different materials forming the slope must be assessed under the loads defined in Stage II.

A given soil type can be treated, in general, as a continuous material and its behaviour will be dependent upon the properties of the material itself. The strength of a rock mass, however, will be dominantly dependent upon the nature of the discontinuities (faults, joints, etc.) that bound the intact rock. Thus to determine the strength/deformation characteristics of the rock mass requires a prior knowledge of its geology (Stage I).

Once these three stages have been adequately delineated, the various parameters are assembled, along with the geometry of the slope, to provide the 'input data'. The analytical techniques employed must successfully represent these parameters and their interaction. Thus, for a stability of a slope to be carried out, a series of assessments must previously be made and brought together in a model so that the degree of stability or instability can be determined. Stability analyses provide the index obtained from the model known as a Factor of Safety. As a result, a comparison can be made of varying slope geometries so that the optimum configuration can be determined compatible with an adequate Factor of Safety.

The importance of relating stability analyses to practical experience must be emphasized since it is through careful examination of the behaviour of rock slopes currently forming the slopes of open pits that a reliable series of design techniques can be established.

METHOD OF ANALYSIS

Once an evaluation of the governing parameters in a slope has commenced, it becomes necessary to develop a model that represents the particular conditions found within the pit. Such models fall into two categories.

1. **Physical Scale Models**

   Such models have been constructed at research establishments and universities, (5). They adequately scale most of the inherent properties of the rock mass. The building and subsequent failing of a scale model is both time consuming and costly. Thus to examine many possible configurations of slopes becomes, in effect, an impossible task. This should not be taken as a reason to inhibit such experiments at universities, etc. since valuable improved understanding can be attained from them that can subsequently be incorporated in mathematical models.

2. **Mathematical Models**

   Mathematical models are the most practical analytical tools available today since they provide, in general, adequate representations of the rock masses forming the slopes, a high degree of flexibility, and relatively low operating costs. Mathematical models fall into two general categories:

   1. Limit Equilibrium Analysis
   2. Stress/Strain Analysis.

Limit Equilibrium Analysis

This type of analysis makes the following assumptions:

(i) that a failure surface can be postulated with a geometry that will allow sliding to take place — a 'kinematically possible' surface.

(ii) that the distribution of stresses in equilibrium along the potential failure surface can be computed utilizing a knowledge of density, water pressure, and seismic loading.

Once the above assumptions are accepted, the analysis computes the shear resistance along the potential failure that would be necessary to bring the material enclosed by the slope profile and the failure surface to the point where movement is about to occur. This condition is known as Limiting Equilibrium.

If the **computed** shear resistance at Limiting Equilibrium is equal to the **measured** available shear resistance, failure is theoretically imminent. If the computed shear resistance is greater than the available shear resistance, failure **should** have taken place. Where the computed shear strength is less than the available shear strength the Factor of Safety will be greater than 1.0. Under such conditions the potential surface will be stable.

The comparison of the available and computed shear resistances is carried out by means of an index known as the Factor of Safety. Thus the Factor of Safety is simply that factor by which those parameters that control the available shear resistances must be divided in order to bring the material to the point of Limiting Equilibrium.

The repeated computation of a Factor of Safety for all the many kinematically possible failure surfaces allows comparisons to be made between each surface. The surface providing the minimum Factor of Safety will be the most dangerous for that particular pit slope geometry.

Stress/Strain Analysis

In the field of material mechanics, this type of analysis was originally restricted to conditions where the material's properties and behaviour could be assumed to be homogeneous. Such conditions are, rarely, if ever, found in the materials forming open pit slopes.

This type of analysis, however, has become much more versatile since the development of powerful numerical analysis techniques that parallelled the development of large digital computers.

The 'Stress/Strain' type of analysis predicts the nature and extent of displacements compatible with equilibrium within the slope, as it is mined out. Some limitations exist in the utilization of this technique, of which the most important one is the difficulty of adequately defining the displacement characteristics in the highly fractured rock mass and/or soil that almost always constitutes the slopes of open pits mines. This difficulty does not preclude the use of this technique, in some circumstances, because an 'educated guess' can be made of the stresses utilizing field information.

In summary, these two techniques are different in that the 'Limit Equilibrium' technique assumes a failure surface and a stress distribution and then locates the surface most likely to allow movement, whereas the 'Stress/Strain' type of analysis assumes an initial stress state and then computes where displacements and potential failure will occur. 'Limit Equilibrium' indicates when total collapse should occur, while the 'Stress/Strain' approach indicates an earlier state when an initial partial and localized failure would occur. Provided the correct assessments have been made about the so-called 'input information' the latter technique will normally be more conservative and produce flatter slopes. In the author's opinion, any pit design based solely on the 'Stress/Strain' type of analysis where homogeneous material properties are assumed would be too conservative, except perhaps where only limited displacements could be tolerated, such as near a skipway or conveyor. However, the 'Stress/Strain' type of analysis can often be usefully employed as a guide to the best locations for displacement monitoring devices (borehole extensometers, surface hubs, etc.) provided intelligent estimates of the material's properties can be made.

THE ESTABLISHMENT OF THE GOVERNING PARAMETERS IN A SLOPE ANALYSIS

This section is a brief summary of what has been covered, in detail, in this symposium, however, it is required to place subsequent aspects of this paper in context.

In the introduction to this paper an outline of the governing parameters, or input data, has been given and the importance of establishing the geological conditions was emphasized.

The 'Limit Equilibrium' type of analysis was shown to require a failure surface, or surfaces, to be selected. This surface should be compatible with the discontinuities that intersect the rock mass.

The 'Stress/Strain' type of analysis selects the area where the rock discontinuities will be displaced. This presupposes that the geology had been modelled adequately. (Further discussion will be given to the problem of geology in a later section of this paper.)

A numerical value for the ability of the rock mass to deform is required for both types of analysis. The accuracy and degree of reliability in assessing this value will directly influence the amount of conservatism called for in the overall evaluation of the stability of a slope.

An examination of the deformation characteristics of the materials forming a slope shows that there are the following types:

1. Soils that initially deform elastically, but subsequently plastically.

2. Soils and rock discontinuities that initially exhibit elastic behaviour but later reach a peak resistance. Once peak resistance is attained, further displacement causes a gradual reduction in resistance until the level of residual resistance is attained, which is the lowest level of resistance that can occur. The manner in which the residual resistance is attained differs markedly between soils and joint surfaces. In the latter case, the initial surface roughness controls failure since rougher surfaces require greater deformation before residual resistance is attained.

3. Rock masses that have closely spaced discontinuities exhibit a high degree of interlock. As a result of interlocking, peak resistance differs only slightly from residual resistance since the individual blocks reorient themselves, which involves rotation and partial fracturing of the intact rock (19) (20).

Induced stresses are associated with the deforming characteristics of materials that compose the slope. Thus for any state of displacement, there is a corresponding state of stress. Once the stress overcomes the available strength, the material may be said to be in a state of failure.

The tangential shear stress and the normal stress applied across the failure surface at failure are the critical parameters in most 'Limit Equilibrium' calculations: A Failure Criterion is a relationship that describes the combination of the two types of stresses.

In Fig. 2 a diagrammatic representation is given of the various forms of Failure Criteria compatible with different slope materials. This will serve as a guide only since some materials can fall into two categories depending on their state and location in the slope.

The concept of a non-linear failure criteria for rocks is important since it is a departure from practice involving soil materials. The non-linear concept, which has been developed by Patton (6), Ladanyi and Archambault (7), and Morgenstern (8), has been slightly modified here into a relatively simple form. The main reasons for adopting the non-linear approach in preference to the linear method are the following:

(i) Test data is usually described more closely by a non-linear failure criterion than by

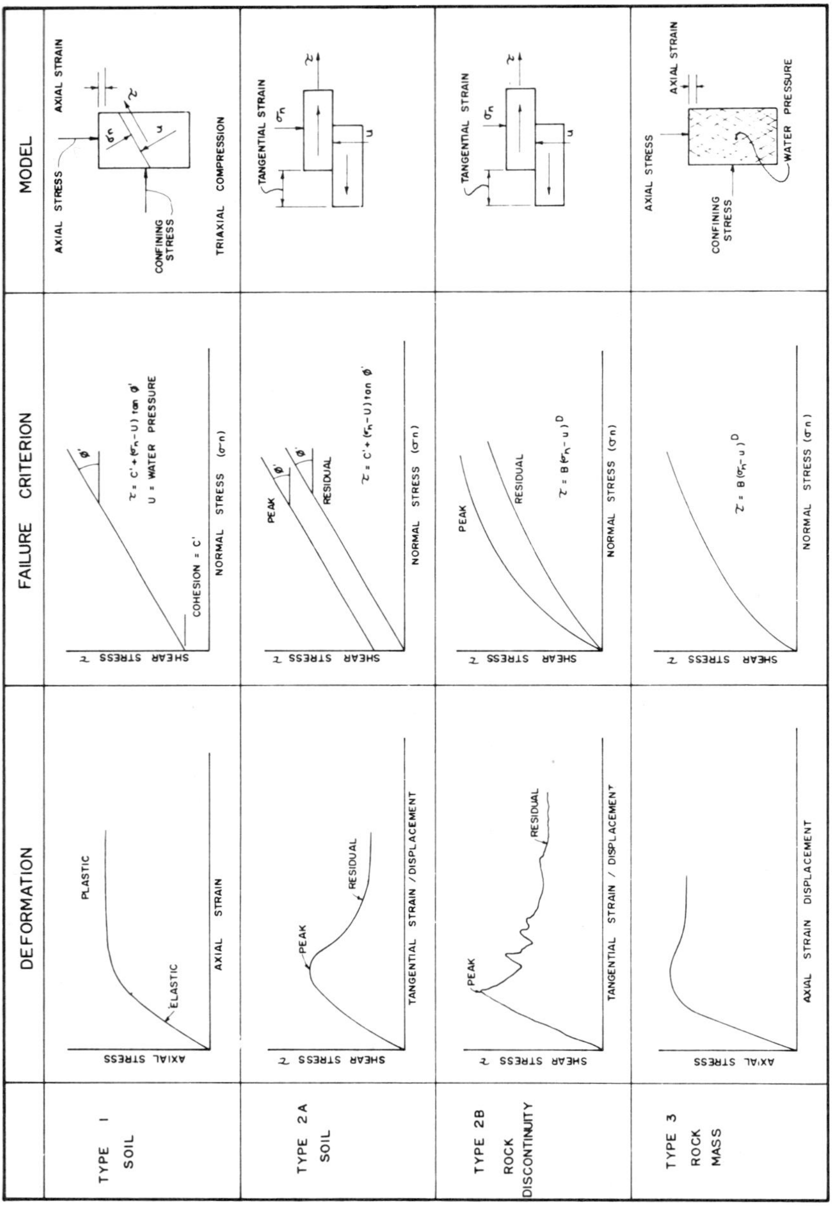

Fig. 2 —Failure criteria for different slope materials.

a linear failure one.

(ii) The effect of small 'errors' in the choice of a value for the cohesion intercept (c') is critical to the computed Factor of Safety if the linear criterion is used.

(iii) The choice of a simple friction coefficient (tan ϕ) (linear criterion) that may be adjusted for the surface roughness of rock joints is often conservative.

(iv) The effect of water pressure — described elsewhere in this conference — can reduce a portion of the rock mass to failure by reducing the effective stress. If a non-linear criterion is used the effect of water will be correctly reflected, at low normal stresses while apparent stability would be maintained if a linear criterion was used.

These important factors are shown on Fig. 3.

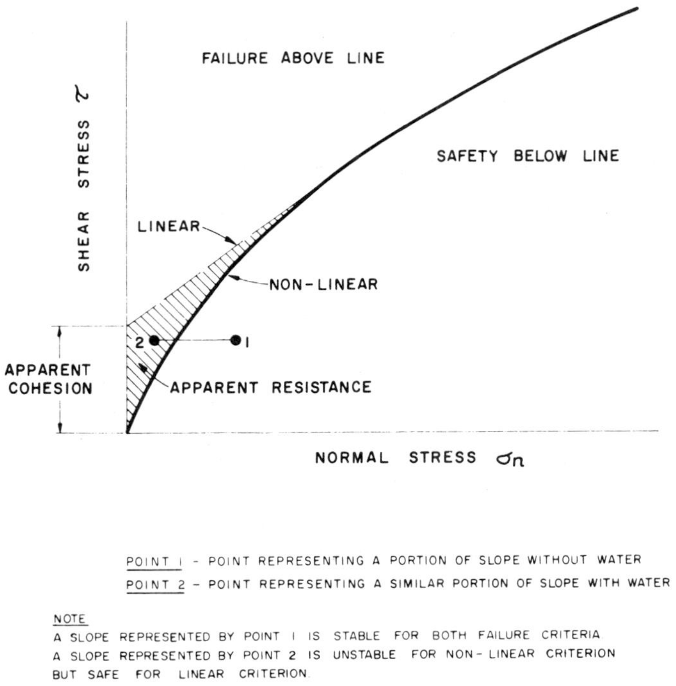

Fig. 3 — Linear and non-linear failure criteria.

SPECIFIC ANALYSES

No single type of mathematical model adequately represents all of the conditions found in pit walls, and therefore many different types have been developed. The divisions are:

(a) Limiting Equilibrium

(i) Inclined wedge

(ii) Cylindrical arc

(iii) Surface of varying curvature

(b) Stress/Strain

Finite Element

The equations and formulation of three main types of mathematical analysis are included in Appendix 1. There is also a description of 'Limiting Equilibrium' type of analysis outlined below.

Inclined Wedge Method

This example involves a simple wedge formed by a single discontinuity (see Fig. 4).

The forces acting down the plane are found by resolving the forces acting on the wedge and may be expressed as the following sum:

$$W \sin \alpha + Wk \cos \alpha + Q \cos \alpha \qquad \ldots\ldots\ldots (1)$$

The normal stress σ_n is found by the resolution of the forces acting normal to the discontinuity.

$$\frac{W \cos \alpha - Wk \sin \alpha - Q \sin \alpha}{L} \qquad \ldots\ldots\ldots (2)$$

The shear resistance for limiting conditions is calculated from:

$$BL\left(\frac{W \cos \alpha - Wk \sin \alpha - Q \sin \alpha}{L} - u\right)^{D} \qquad \ldots\ldots\ldots (3)$$

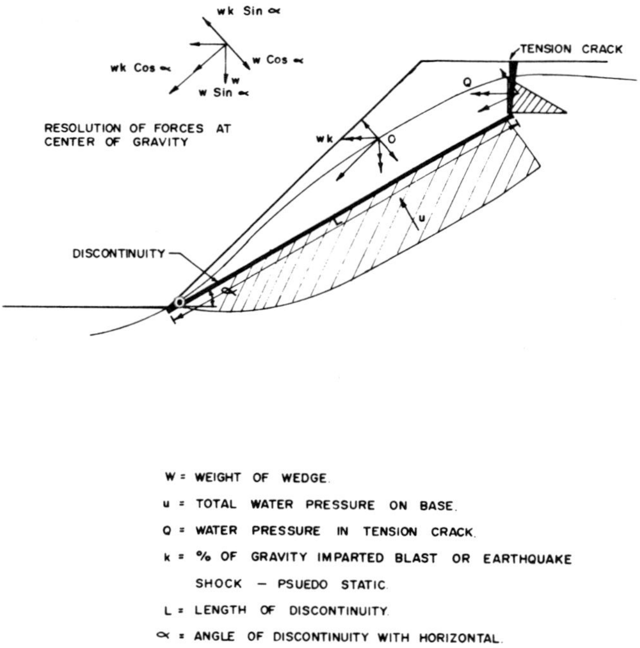

Fig. 4 — Model for failure of an inclined wedge.

where B and D are known (see Fig. 2). However, since a definition of the Factor of Safety (F) is implied in a Failure Criteria, the Factor of Safety may be computed from the following:

$$F = \frac{BL\left(\frac{W\cos\alpha - Wk\sin\alpha - Q\sin\alpha - u}{L}\right)^{D}}{W\sin\alpha + Wk\cos\alpha + Q\cos\alpha} \quad \ldots\ldots\ldots (4)$$

If the above problem is simplified as follows:

(i) no water is present ($\bar{u}$ = 0)

(ii) no psuedo static seismic shock (k = 0)

(iii) no tension crack (Q = 0)

(iv) the failure criterion is linear (D = 1.0 and B = tan ϕ)

the Factor of Safety can be reduced to:

$$F = \frac{\tan\phi'}{\tan\alpha} \quad \ldots\ldots\ldots (5)$$

Although the model is relatively simple, the principles involved are important and are identical to more complex analyses of the 'Limit Equilibrium' type.

The Cylindrical Arc Method

This method presupposes that the material forming a pit slope will fail along a surface which in two dimensions is a circular arc. Such an assumption is often valid in soils since experience has shown that failure can take place in this manner. The technique of dividing the slope into vertical slices enables each section's contribution to the total resistance of the surface to be computed. This type of analysis can also be used reasonably accurately for rock types where the spacing between the discontinuities is very small compared to the total dimensions of the slope and where the spatial orientation of the discontinuities is approximately evenly distributed.

Surface of Varying Curvature or Non-Circular Arc Method

This method is the general form of the 'Limit Equilibrium' analysis. The most advanced form of it was developed by Morgenstern and Price (13) whose method is capable of dealing with any shape of failure surface and any failure criteria. However, Janbu's method (12) gives an approximate solution that is normally adequate, particularly when the potential failure surface is shallow and near the pit face.

Finite Element Methods

These methods are a 'Stress/Strain' type of analysis that involve dividing the pit slope walls into a mesh of triangular or rectangular elements. The corners or nodes of these elements are connected so that similar displacements will occur. Each element is assumed to be homogeneous. A typical mesh is shown on Fig. 5.

With the aid of a digital computer, the displacement in two directions is computed. The calculations are based on equilibrium between the external loads due to self-weight, water pressure, etc., and the forces induced by the deformation of each element.

This method is becoming increasingly effective since joints and faults can be modelled (9). However, at present there is a limit to the number of such features that can be allowed for.

The displacements can be related to a failure criterion and if they exceed a critical value, for a particular section, that portion of the rock will have reduced strength. Therefore by using

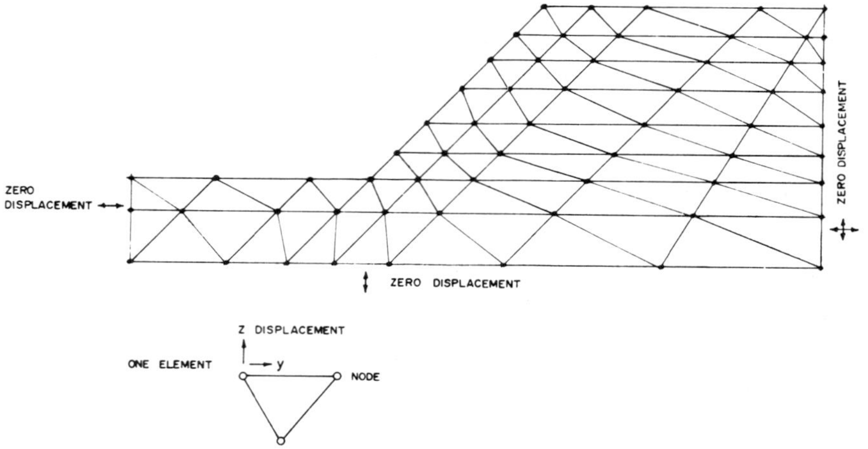

Fig. 5 — Typical finite element mesh for a slope.

a series of values in the analysis, a simple form of progressive failure can be modelled.

TYPICAL RESULTS

A simple comparison of the effects of varying the shear strength parameters is shown below. A slope of 70° with a cut height of 100 ft. gave the following results for the cylindrical arc method.

COHESION (c′) lb./sq. ft.	FRICTION COEFFICIENT (Tan ϕ' or B)	FACTOR OF SAFETY (F)
1000	0.63	1.61
0	0.63	1.05

In this simple example, it is clear that the effect of cohesion is critical to a theoretical stability analysis. This effect is far less marked for higher slopes. This problem emphasizes the need to define the limits of normal stress for which a mathematical criterion of failure is valid.

As a comparison to the above analysis, a finite element analysis has been carried out for a similar slope and is shown on Fig. 6. The results of this analysis were obtained using a 'Stress/Strain' modulus of 1 million p.s.i. and are comparable to the rock mass values quoted above. The results indicate that the maximum displacements occur near the toe and are of the approximate order of 15 in. horizontal movement and 3 to 5 in. vertical movement. In this case, the slope has been theoretically mined and the unloading of the slope has contributed to the displacements. The largest displacements occur near the surface and are reduced further into the rock mass. For example, at a point 40 ft. from the face and midway up the slope, the displacements are 3 and 7 in. for the vertical and horizontal directions respectively. These displacements would be enough to fail most rock joints and this analysis is therefore more conservative than the 'Limit Equilibrium' method of analysis.

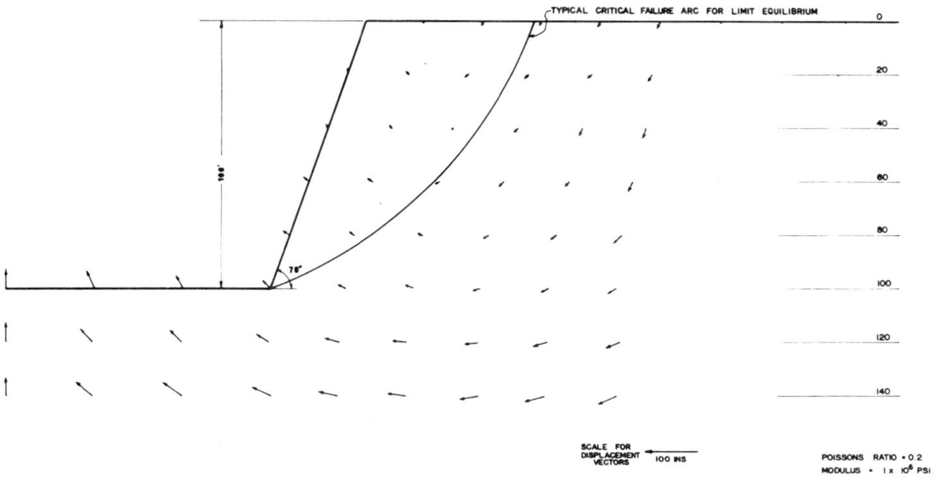

Fig. 6 — Resultant displacements for finite element analysis.

LOGICAL USE OF STRUCTURAL DATA

If the 'Limit Equilibrium' approach is adopted, the choice of the failure surface becomes essential.

Most operators of open pits will have noticed that failures involving one or more benches have occurred in the shape of a wedge. The slope and the crest form two of the faces, with the other two being formed by discontinuities such as joints or faults intersecting the slope.

This simple case can be analyzed utilizing simple vector algebra (Appendix I). The author has computerized this analysis to speed calculation since the rock mass forming an open pit slope contains many structural discontinuities. This information is presented in the form of a stereo projection. Fig. 7 shows how a plane and the normal to the plane and also the intersection of two planes are projected on the hemispherical plane of the reference sphere. If the discontinuities were properly measured by the geologist, a statistical model of the spatial orientations of the discontinuities intersecting the rock mass is presented by the stereo projection.

A certain number of wedges can be formed by combining various pairs of discontinuities. The total number of possible combinations is $\frac{(n.n-1)}{2}$ if 'n' is the number of discontinuities. This means that for 500 joint measurements of dip and dip direction, there are 124,750 possible combinations. Fortunately, however, a number of wedges cannot fail due to their particular geometry in relationship to the slope (see Fig. 8).

A simple check can be made utilizing the stereographic projection to determine if a particular wedge has a geometry that precludes failure. Traces are drawn of the slope and the two discontinuities and if the intersection of the two discontinuities falls outside the slope trace (i.e. flatter) then failure can **potentially** take place.

A stability analysis is carried out on each of the wedges that can potentially fail. The principles outlined previously are employed. In this case the analysis is truly three dimensional.

For a given geological sample of structural discontinuities, there is a total number of possible combinations (C). For a given mode of failure (i.e. linear sliding, rotational sliding or toppling), there are a number of potential possible failures (S).

In some cases, some of the potential possible failures will overcome a predetermined resistance and fail.

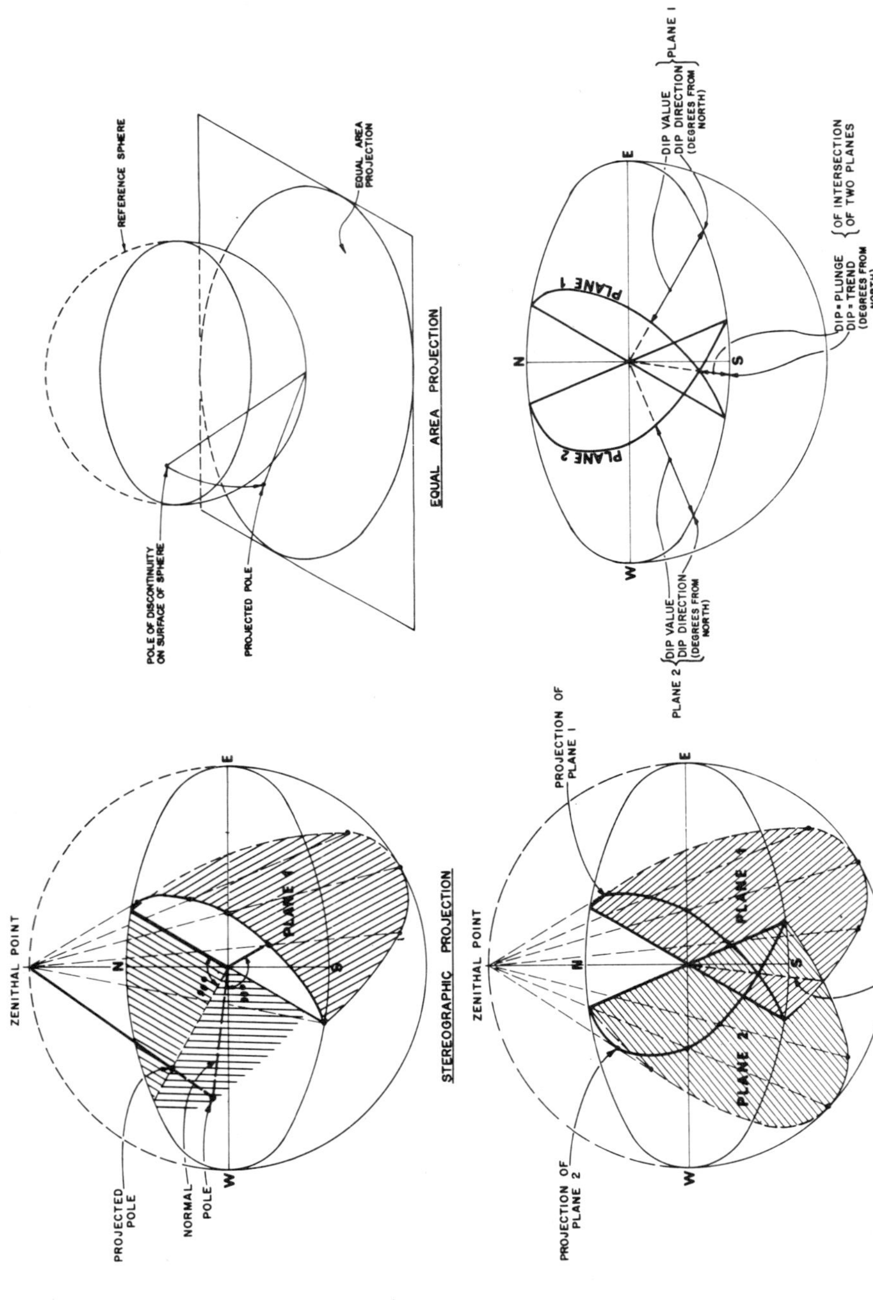

Fig. 7 — Reference sphere projections.

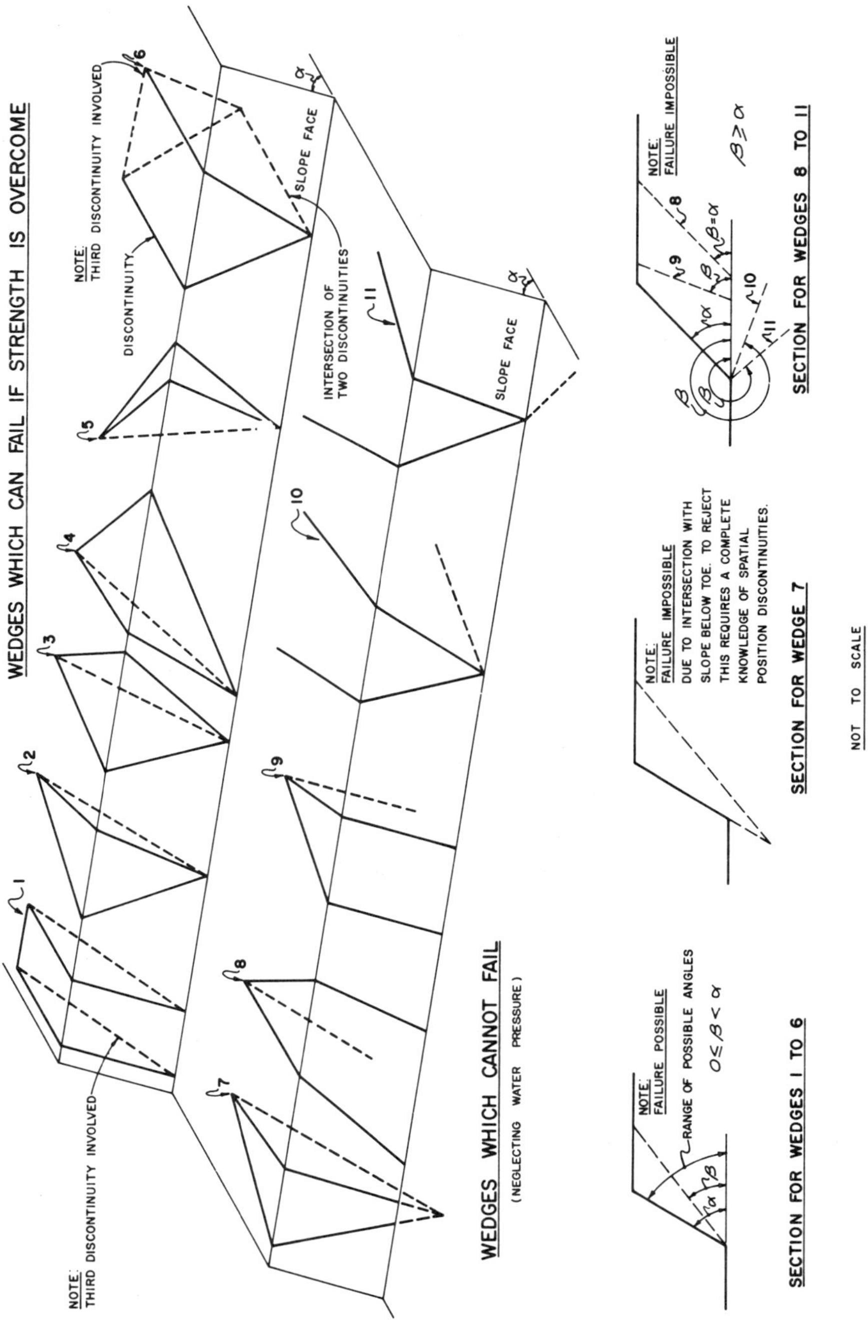

Fig. 8 — Typical wedges in rock slopes.

An example, taken from a section of an open pit, is included to demonstrate this type of analysis. Only the case for linear study is discussed since the other modes of failure are less critical. The initial data for some 188 discontinuities is shown in the equal area projections in Fig. 9.

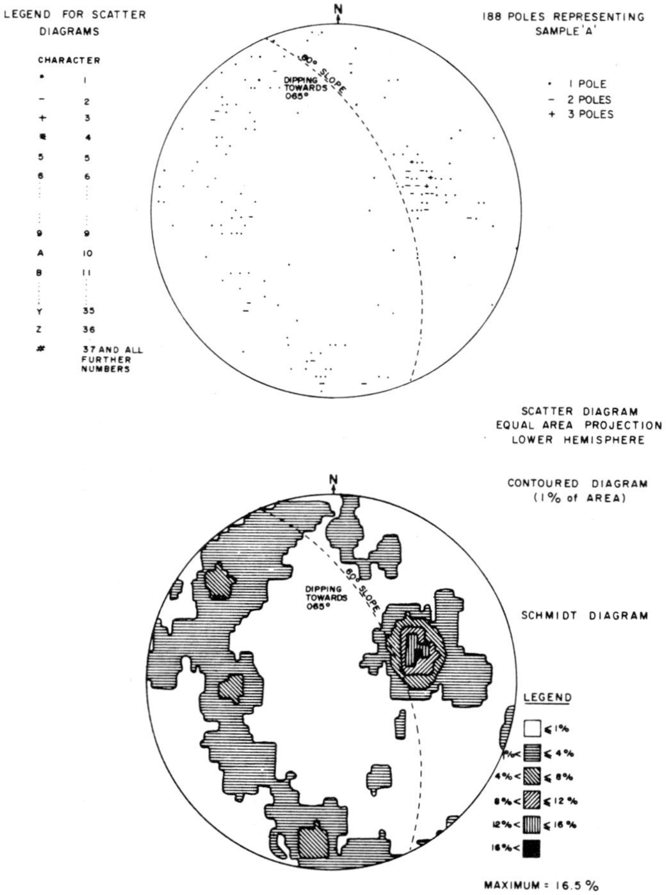

Fig. 9 — Typical equal area projections of a sample 'A' of joints from an open pit.

A slope dip direction of 065° was examined for slope angles of 40°, 50°, and 60°.

The histogram of the Factors of Safety is shown in Fig. 10 for the different slope angles and for two apparent friction angles of 50° and 60°. The sensitivity of the analysis to varying strengths and geometries is demonstrated.

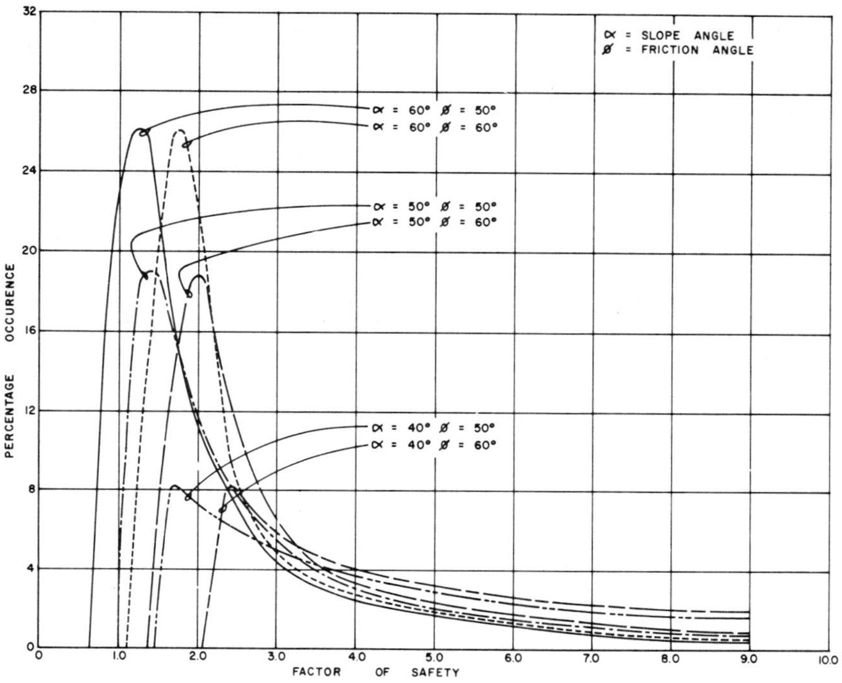

Fig. 10 — Histogram of the Factors of Safety for sample 'A' for varying slopes and strengths.

The statistics of these analyses, for 50° slopes dipping towards 336° and 156° are shown below:

SUMMARY OF TYPICAL WEDGE ANALYSIS

Original Data = 188 discontinuities

Total number of combinations (C) = 17,578

Mode of Failure — Linear Sliding

Slope Dip Direction (deg.)	065						336		156	
Slopes Angles (deg.)	40	50	60	40	50	60	50	50	50	50
Friction Angle (deg.)	50			60			40	50	40	50
Minimum Factor of Safety	1.50	1.02	0.56	2.18	1.49	1.17	0.72	1.02	0.76	1.07
Value of S	1796	3389	4814	1796	3389	4814	3745	3745	4259	4259
Value of T	0	0	237	0	0	0	137	0	77	0

The possibility or likelihood of potential failure wedges being formed varies as the attitude of the slope face is changed with respect to a particular structural sample. The analysis makes the assumption that the location in space of any two discontinuities is such that they can intersect the slope face at the toe. This is a conservative assumption, and the analysis assumes that the structural data is a representative sample of the rock mass.

The example, a 50° slope dipping towards 065° and a friction angle of 60° is shown in greater detail in Figs. 11 and 12. The intersections of the potential failures (S) are shown in the upper diagram as a plot of individual poles utilizing an equal area projection and in the lower diagram as a Schmidt contoured diagram of the equal area plot. Fig. 12 shows the same type of plots for those possibilities which can fail.

This type of analysis has been applied to pit slopes in general, and particularly to major haul roads, where it is necessary to locate sections that require reinforcement (drainage, etc.). This technique is currently being expanded to take into account a third joint set.

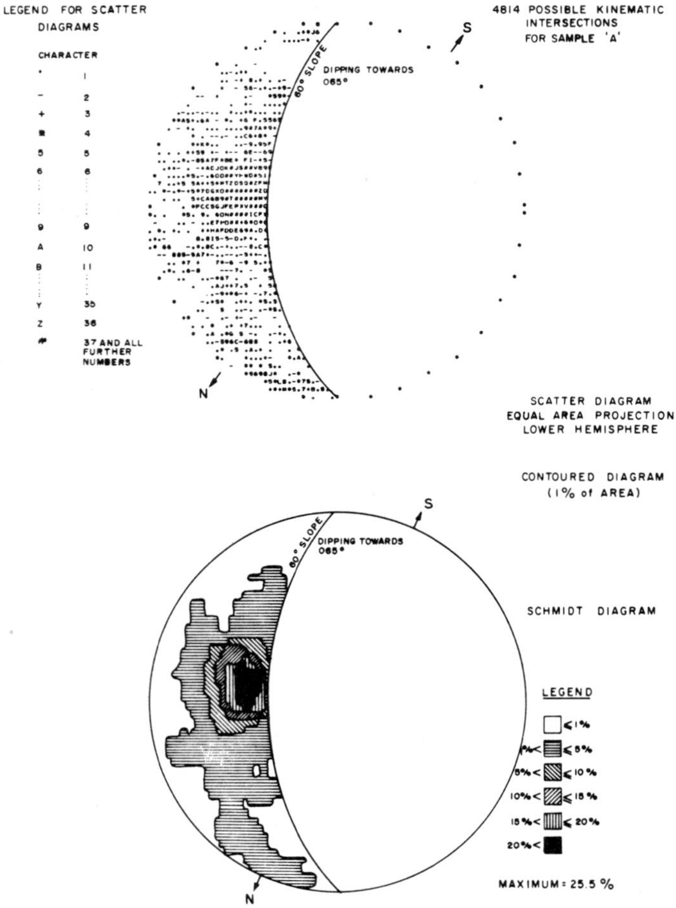

Fig. 11 — Projections of all possible kinematic intersections for sample 'A'.

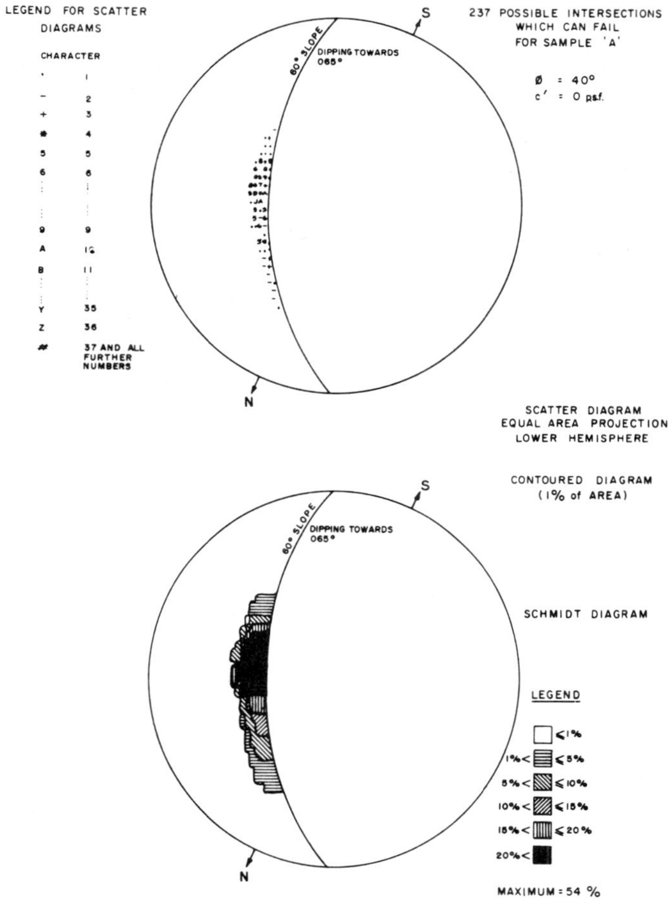

Fig. 12 — Projections of all possible intersections for sample 'A' which will theoretically fail.

CONCLUDING REMARKS

This paper has described several powerful analytical procedures that have been developed. In many cases, these procedures have outstripped the input information.

Experience has shown that it is **essential** to measure and monitor slopes before and during a mining operation. The open pit is itself a model that consistently provides valid information. If this information is continuously monitored and analyzed, the pit management will be able to arrive at an integrated mining strategy.

These requirements are not academic but simply practical. There is a place for laboratory experiments, but in situ behaviour is what is really required to provide a realistic assessment of pit slope stability. For feasibility studies, a range of alternatives should be provided. These can be obtained by carefully utilizing exploration holes and adits with the result that 'ball park' answers can be arrived at.

Bench failures occur continuously in most mines. In some cases, these may only be blocks of tens of feet in dimension. Such cases are much better than any laboratory experiment if care is taken to measure the displacement over a period of time. The failure of a section of Chuquicamata Open Pit (10) is a classical example of an instrumented failure. A 'back analysis' of such a failure will enormously increase the amount of knowledge about the future behaviour of the pit slopes. 'Back analysis' means, in the case of 'Limit Equilibrium', analysis, assessing the shear strength parameters that must have prevailed when sliding took place. In order to carry this out, the following parameters are required.

(i) A geological description of the rock mass, and particularly of the failure plane.

(ii) A knowledge of any water pressures present.

(iii) A knowledge of the displacement characteristics with time.

If a 'Stress/Strain' analysis is used, the real displacements can be compared with those predicted using various modulus characteristics. When agreement is obtained the correct modulus of the in situ rock mass is established.

APPENDIX I

MODEL FOR THE CLASSICAL CIRCULAR FAILURE

This model has been adapted by Bishop (11) for the more general failure criterion.

The potential failure surface is divided into many integral vertical sided slices with two triangles at the crest and the toe.

All the slices, total number n, are arranged so that the sides of each slice coincide with any lithological or structural change. The base of each slice is theoretically an arc of a circle but is practically regarded as a straight line. Therefore, greater accuracy is obtained if the slice width is small compared to the total mass. A diagram of a typical slice is shown on Fig. 1-1.

At the limiting condition, the sum of the resistance moments must equal the sum of the driving moments, if rotational equilibrium is to be satisfied.

$$\text{Thus } \frac{R}{F} \sum_{1}^{n} s = \Sigma\, W(R \sin \alpha + kd) \quad \text{.........(1)}$$

Where s = the particular shear resistance for the material through which the base of each slice passes when the material is at the point of failure.

d = Seismic moment arm for each slice about the centre of rotation.

All the other symbols are described in the figure.

In equation (1) by substituting the value of the shear resistance (s) the following equation is produced:

$$F = \frac{R}{\sum_{1}^{n} W(R \sin \alpha + kd)} \sum_{1}^{n} At + B(P_n)^{D}/(t)^{D-1} \quad \text{.........(2)}$$

Where $P_n = t(\sigma_n - u)$ = the effective normal force acting on the base of each slice i.e. the failure surface.

The shear strength criterion used is:

$$\tau = A + B(\sigma_n - u)^D$$

A = Shear strength intercept at zero normal stress

B = Friction factor

D = Exponent for power law failure criteria

NOTE: If D = 1.0

Coulomb law is obtained, i.e.

$$\tau = c' + (\sigma_n - u) \tan \phi'$$

c' = cohesion under effective stress conditions

ϕ' = effective friction angle under effective stress conditions

In addition to the moment equation (1), the components of forces acting vertically are resolved for each slice. This provides the quadratic equation from which the effective normal force can be solved.

$$P_n^D (B \sin \alpha / F (t)^{D-1}) + P_n \cos \alpha + \quad \ldots$$

$$\ldots \; Ab \tan \alpha / F - W(1 - r_u + k \sin \psi) - (V_m + V_{m+1}) = 0 \qquad \ldots\ldots\ldots (3)$$

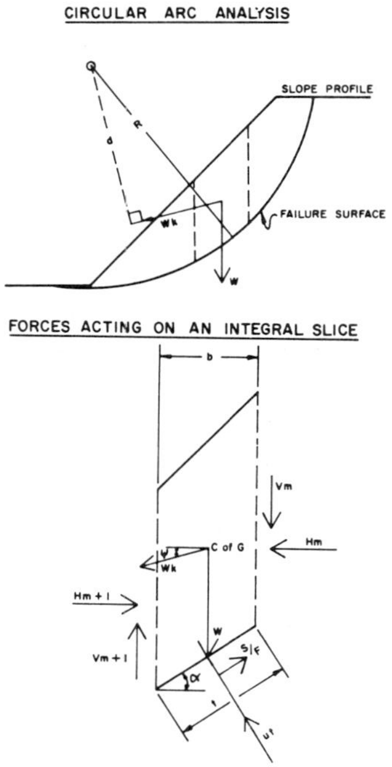

Fig. 1-1 — Cylindrical arc analysis.

The total effect of the shear forces (V_m, V_{m+1})must be zero, and if the term($V_{m+}V_{m+1}$) is neglected, only a small error is involved.

Equation (3) can be solved by use of the Newton's numerical technique:

$$P_n = P_{n-1}) - \frac{f(P_{n-1})}{f'(P_{n-1})} \qquad \text{........ (4)}$$

The process involves an initial estimate of the effective normal force, P_{n-1}. This is substituted in the following equation:

$$Pn = \frac{\dfrac{(D-1)(P_{n-1})^{D}\ B\sin\alpha}{F(b\sec\alpha)^{D-1}} - \dfrac{Ab\tan\alpha}{F} - W(1-r_u + k\sin\psi) - (V_m + V_{m+1})}{\dfrac{D(P_{n-1})^{D-1}\ B\sin\alpha}{F(b\sec\alpha)^{D-1}} + \cos\alpha} \qquad \text{........ (5)}$$

From equation (4), a new and better estimate is produced (P_n). This process is repeated until the difference between the estimated and the computed value is negligible. In this manner, all the effective normal forces acting across the base of each slice are computed and substituted into equation (2). This equation then gives the overall Factor of Safety. However, it will have been noted that an initial value of the Factor of Safety is necessary to solve equation (5). Thus it is necessary to repeat the whole process again, i.e. solve equations (5) and (2) with the improved estimate of the Factor of Safety. This outer 'loop' of computations is repeated until the difference between the estimated and computed Factor of Safety is negligible.

In this manner the Factor of Safety for a particular failure surface can be computed. Relocation of the critical failure surface, by changing the centre and radius of the arc, enables the minimum Factor of Safety of a particular slope to be computed.

This apparently complex manipulation can be easily achieved with the aid of a computer. (Some 500 potential failure surfaces can be examined in ten minutes utilizing an IBM 360-50).

NON-CIRCULAR FAILURE

The type of analysis developed by Janbu (12) is an approximate solution. (see Fig. 1-2). The rigorous solution has been provided by Morgenstern (13). The slope is divided into slices (n) and the forces that resist sliding are equated with those causing sliding to take place.

The equation delineating the Factor of Safety (F) is:

$$F = \frac{fo\,\Sigma_n\,[c'b + (\triangle W - Ub)\tan\phi'] / \cos^2\alpha\,(1+\tan(\alpha/F)\tan\phi')}{\Sigma_n W\tan\alpha} \qquad \text{........ (6)}$$

This utilizes a correction factor (fo) based on the geometry of the slide that allows the forces acting between the slices to be balanced.

The Factor of Safety is initially assumed and then the computed result is placed back into the equation with the process being repeated until convergence occurs.

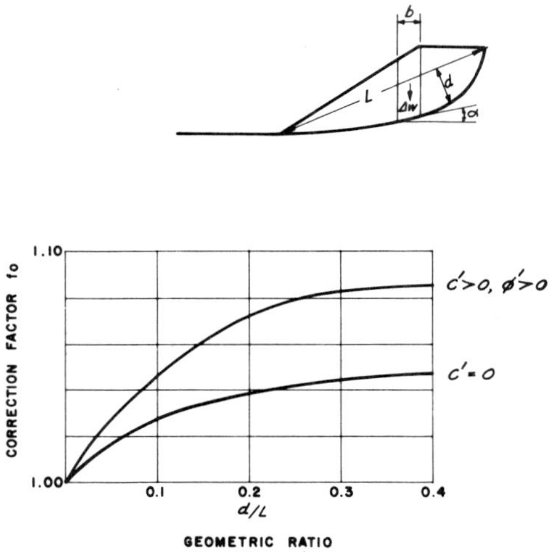

Fig. 1-2 — Janbu's analysis.

WEDGE ANALYSIS

This type of analysis involves the use of vector algebra. Wittke (14), John (15), Londe et al (16), and Goodman (17) developed this technique which has been computerized and made more general by Pentz, (18).

Each discontinuity is represented by the vectors V1,2 and W1,2 which represent the lines of maximum dip direction and the normals to the discontinuities respectively (see Fig. 1-3).

The analysis examines the kinematics of three types of failures:

(i) Linear sliding on
either a) both planes
or b) either one plane or the other.

(ii) Rotational sliding

(iii) Toppling.

The latter two cases are beyond the scope of this paper, but the equations for the Factors of Safety are given for the first type of failure.

Sliding on plane 1 OR plane 2

$$F = \frac{AR_{1'2}\{A + B\,[\pm \dfrac{\vec{W}_{1'2}\cdot(\vec{WT} + \vec{SV}_{1'2})}{AR_{1'2}} - PW_{1'2}]^{D}}{\vec{V}_{1'2}\cdot(\vec{WT} + \vec{SV}_{1'2})} \qquad \ldots\ldots\ldots (7)$$

Sliding Both Planes Parallel to Intersection Line

$$F = \frac{AR_{1'2}\ \{A + B\,[\dfrac{\vec{N}_{1'2}}{AR_{1'2}} - PW_{1'2}]^D}{[\vec{M}]} \qquad \ldots\ldots\ldots (8)$$

Where $\vec{M} = \vec{X} \cdot \dfrac{\vec{X}}{|X|} (\vec{SX} + \vec{WT})$

Where $AR_{1'2}$ = Area of two planes
$\vec{WT}$ = Self-weight vector
$\vec{N}_{1'2}$ = Component of normal force on each plane
$\vec{SV}_{1'2}$ = Seismic vector parallel to either plane
$\vec{SX}$ = Seismic vector parallel to the intersection line
$\vec{PW}_{1'2}$ = Pore water pressure
$\vec{X}$ = Vector describing the intersection line

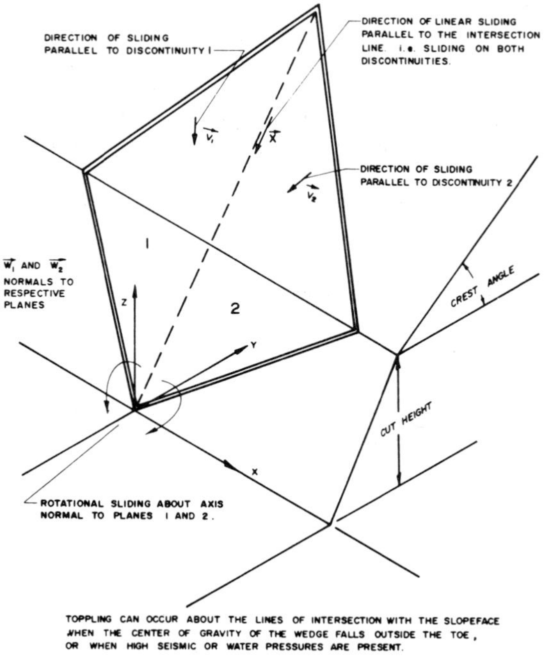

Fig. 1-3 — Wedge analysis.

REFERENCES

1. Broadbent, C.D. and Rippere, K.H., "Fracture Studies at Kimberly Pit" Open Pit Mining Symp., Johannesburg, 1970.

2. Robertson, A. MacG., "The Interpretation of Geological Factors for Use in Slope Theory", Open Pit Mining Symp., Johannesburg, 1970.

3. Rosengren, K.J., "The Rock Mechanics of the Black Star Open Cut, Mount Isa", Ph.D. Thesis, Australian National University, Canberra, 1968.

4. Terzaghi, R.D., "Sources of Error in Joint Surveys" Géotechnique, Vol. 15, 1965, pp. 287-204.

5. Barton, N., Ph.D. thesis in preparation, Imperial College, London University.

6. Patton, F.D., "Multiple Modes of Shear Failure in Rock and Related Materials," Ph.D. Thesis, University of Illinois, 1966.

7. Ladanyi, B. and Archambault, G., "Evaluation of Shear Behaviour of a Jointed Rock Mass," Proc. 11th Symp. on Rock Mechanics, Berkeley, 1969.

8. Morgenstern, N.R., "Ultimate Behaviour of Rock Structures", Ch. 10., Rock Mechanics in Engineering Practice, ed. by K.G. Stagg J. Wiley, New York, 1968.

9. Goodman, R.E. and Taylor, R.I., "Methods of Analysis for Rock Slopes and Abutments, A Review of Recent Developments," Proc. 8th Symp. on Rock Mechanics, 1966, pp. 303-320.

10. Kennedy, B.A., Niermeyer, K.E. and Fahm, B.A., "Major Slope Failure at the Chuquicamata Mine, Chile," Mining Engineering, A.I.M.E., V21 W12, 1969.

11. Bishop, A.W., "The Use of the Slip Circle in the Stability Analysis of Slopes," Géotechnique, Vol. 5, 1955, pp. 7-17.

12. Janbu, N., "Application of Composite Slip Circles for Stability Analysis," Proc. Europ. Conf. on Stability of Earth Slopes, Vol. 3, Stockholm, 1954, pp. 43-49.

13. Morgenstern, N.R. and Price, V.F., "The Analysis of the Stability of General Slip Surfaces," Géotechnique, Vol. 15, 1965, pp. 79-93.

14. Wittke, W., "Method of Analyzing the Stability of Rock Slopes With and Without Additional Loading," Felsmechanik und Ingenieurgeologie, Supplementum II, 1965, pp. 52-79.

15. John, K.W., "Graphical Stability Analysis of Slopes in Jointed Rock," J. Soil Mech and Foundation Div., A.S.C.E., Vol. 94, No. SM2, 1968, pp. 497-526.

16. Londe, P., Vigier, G., Vormerginer, R., "Stability of Rock Slopes, a Three-Dimensional Study," J. Soil Mech and Foundation Div., A.S.C.E., Vol. 95, No. SMI, 1969, pp. 235-262.

17. Goodman, R.E., Discussion, "Graphical Stability Analysis of Slopes in Rock" J. Soil Mech and Foundation Div., A.S.C.E., Vol. 95, No. SM2, 1969, pp. 698-702.

18. Pentz, D.L., Ph.D. thesis in preparation, Imperial College, London University.

19. Rosengren, K.J. and Jaeger, J.C., "The Mechanical Properties of an Interlocked Low-Porosity Aggregate," Géotechnique, Vol. 18, 1968, pp. 317-326.

20. Jaeger, J.C., "The Behaviour of Closely Jointed Rock," Proc. 11th Symposium on Rock Mechanics, Berkeley, 1969.

Chapter 9

FIELD INSTRUMENTATION FOR ROCK SLOPES

by

K. Barron, D. G. F. Hedley and D. F. Coates

Mines Branch, Dept. of Mines, Energy and Resources, Canada.

INTRODUCTION

Preceding papers presented in this symposium have considered the influence of various parameters, such as geologic structure, groundwater, etc., on the stability of rock slopes and have given some analyses and guide lines for slope design. To carry out such analyses it is necessary to put numbers to these parameters; where possible, this is best done on the basis of in-situ observation and measurement. Since none of the design analyses can be considered exact and since conditions vary during excavation, it is also desirable to monitor the behaviour of the slopes as the excavation proceeds. In this manner a feedback of information on the actual excavation conditions can allow a continuous reassessment of the design parameters to be made and can give early warning of incipient failures.

The purpose of instrumenting rock slopes is thus to provide information which fits into one of two categories:

(a) To give basic data to assist on-site assessment and design of the rock excavation.

(b) To provide data for the control and assessment of safety during excavation.

The purpose of this paper is to review briefly some of the instruments and measuring techniques available for gathering this data and to indicate where such data may prove useful. This is not a comprehensive review of all available instrumentation but, rather, is a survey of some of the instruments and techniques with which the authors have had some experience and which have proven useful.

STRUCTURAL ASSESSMENT

Whilst most structural assessment of a site is done on the basis of geologic mapping and core examination, there exists some instrumentation which can, in particular circumstances, assist in this assessment. Seismic instruments can be used to determine the geometry of formations and to give an indication of the relative competency of rocks. These instruments operate using seismic reflection or refraction techniques and, although mainly confined to shallow subsurface exploration, can be useful in distinguishing between soil/bedrock contacts and, in some cases, the extent of weathered surface rock.

A typical survey of this nature is run by placing a geophone on the surface and striking the surface at intervals of, say, 10 ft. along a straight line away from the geophone. The apparatus records the transit time of the first seismic pulse to be picked up by the geophone. Close to the geophone this first arrival pulse has travelled at relatively slow velocity through the topsoil. Further away from the geophone the first arrival pulse will have travelled through the topsoil to the bedrock, through the bedrock (at a much higher velocity) and up through the topsoil to the geophone. The arrival time is plotted against distance of the source from the geophone giving a graph whose slope represents velocity. Points at which the slope of this line changes (i.e., the velocity changes) indicate the depth below surface of the bedrock.

On occasion this type of measurement has been used to indicate the "rippability" of rock, i.e., whether or not mechanical rippers may be used for excavation rather than drilling and blasting. For example, Fig. 1 shows a chart of rippability prepared for one site for one particular combination of mechanical excavation equipment (1).

At greater depths sonic-velocity measurements between boreholes can assist in the extrapolation of core data between boreholes and will sometimes allow a distinction to be made between fractured and competent hard rocks. The sonic velocity is significantly higher in competent rocks than in the same broken rock, e.g., at one site a weathered meta-diabase yielded a sonic velocity of 2000 ft/sec whereas fresh, hard meta-diabase at the same site had a sonic velocity of 14,000 ft/sec. Soils and clays have the lowest velocities, usually in a range of 1000-1500 ft/sec whereas competent quartzites have velocities exceeding 20,000 ft/sec.

Visual Logging

As an assistance on the interpretation of drill core it is sometimes possible, if the expense is warranted, to look at the inside of a borehole for evidence of fracturing, joints and joint orien-

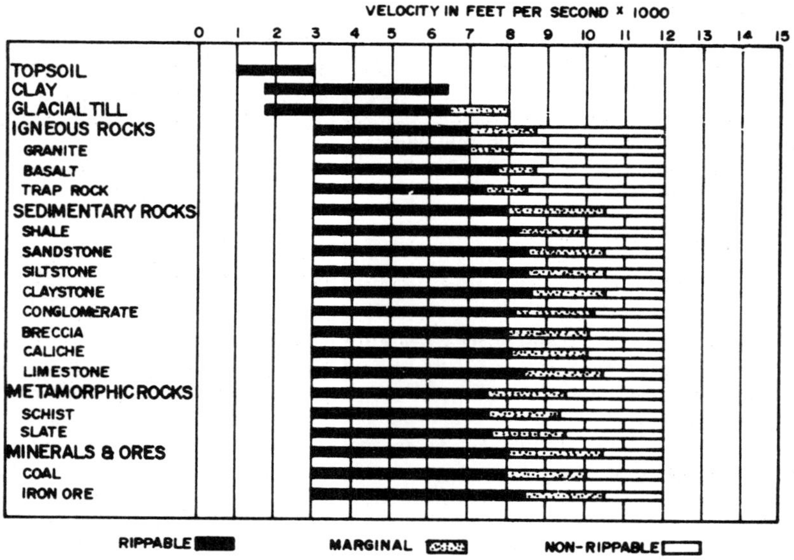

Fig. 1 —Correlation between rippability and seismic velocity (1).

tation, soft ground, caving, voids, etc. The simplest viewing device is the periscope which comprises an inclined mirror and light source, viewed by telescope from outside the hole. These, however, are seldom satisfactory beyond a depth of 20 ft. More elaborate, and expensive, borehole cameras are now available which allow both viewing and photographing of the inside of a 2¼-in. borehole to depths of over 100 ft. With this type of device, the strike and dip of the observed planes may be measured provided care is taken to eliminate slack at the joints of the sections of the insertion rods.

Borehole television cameras provide the ultimate in luxury. These have been used to depths of 400 ft. or more but require a relatively large diameter borehole. One model requires an NX (3-in. diam.) hole whilst another requires a 2½-in. diam. hole. Again measurements of the dip and strike of observed planes may be made if slack in the insertion rods is eliminated. Photographs can be taken from the monitor screen.

IN-SITU PROPERTIES

In-Situ Dynamic Properties

Provided that there is good assurance that the rock is competent, then sonic velocity measurements may be interpreted to yield the elastic dynamic properties of the rock using the following equations:

$$Vs^2 = \frac{E}{2\rho(1+\upsilon)} \text{ and } Vc^2 = \frac{E(1-\upsilon)}{\rho(1+\upsilon)(1-2\upsilon)} \qquad \ldots\ldots\ldots (1)$$

where Vs is the shear wave velocity, Vc is the dilatational wave velocity, E is Young's modulus, υ is Poisson's ratio and ρ is the rock density. In competent rock these results compare favourably with those values obtained from laboratory rock substance tests under uniaxial compression. Thus the primary use of such results would be to assist in identification and classification of the rock substance at the site.

Borehole Penetrometer

This device, developed by the U.S. Bureau of Mines (2), has a hardened pin which may be pumped hydraulically to indent the side of the borehole. This indentation test is capable of distinguishing between hard and soft layers of rock. This can be particularly important in slope assessment where the presence of thin layers of soft rock or of clay in otherwise competent rocks could have a great influence on the slope stability. Identification of such layers if therefore of prime importance.

Plate Load Testing

Plate load tests are carried out in situ to derive information on the rock mass properties, as opposed to the rock substance properties. The basis of the test is that if a plate is placed on a rock surface and the contact pressure is increased the plate will follow the rock deformation and, ultimately, the rock will fail. Test results are then interpreted using elastic deformation theory to determine the equivalent deformation modulus and, using modified bearing-capacity theory, to estimate fundamental strength parameters.

Figs. 2a and 2b illustrate two test arrangements. In the first case the plate is loaded with a jack, the reaction load being taken by an opposite wall of the excavation. In the second case, a more common situation for pit slopes, a suitable abutment for the reaction load is unavailable. In such a case the plate is loaded by tensioning a bolt or cable which is grouted at depth in the rock and which passes through the plate. The plate displacement under load can be measured with a dial gauge anchored to a stable point; in addition, the surface deformation of the surrounding rock may be measured with dial gauges.

(a) Deformation Modulus

If it is assumed that the bearing plate is circular with radius R (if square, use the circle

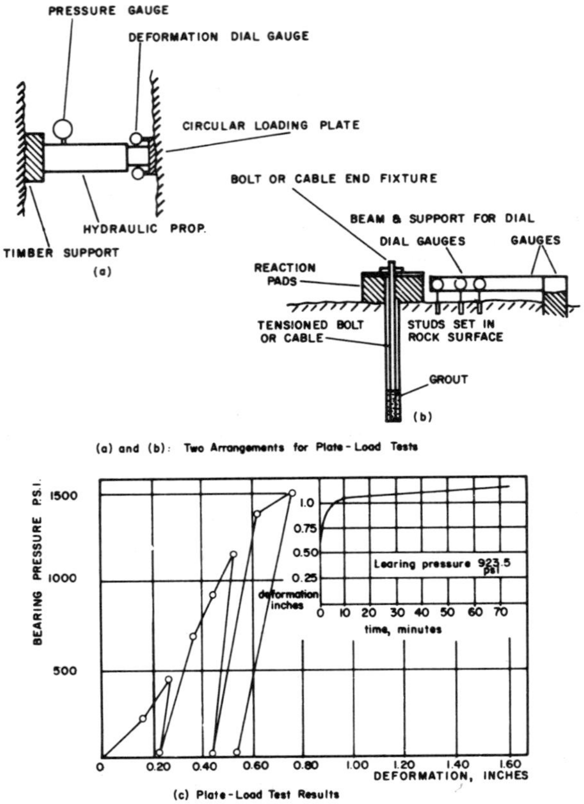

Fig. 2 —Plate load testing.

of equivalent area), that this plate is rigid and that υ is Poisson's ratio, then the displacement of the surface, d, at any radius r, (r $\geqslant$ R), is given by:

$$d = \frac{Q(1-\upsilon^2)}{\pi RE} \sin^{-1} R/r \qquad \dots\dots\dots (2)$$

where E is the deformation modulus and Q is the applied load. Hence, assuming $\upsilon = 0.33$, say, E can be calculated from the load-displacement graph for each point at radius r. In practice several load cycles should be carried out as the first load cycle will include a considerable influence for irrecoverable displacement due to the closing of fissures, etc. For example: plate load tests on the granite hanging wall of an open pit gave E = $(1.30 \pm 0.61) \times 10^5$ psi from a first loading cycle and, from the mean of the second and third cycles, yielded E = $(2.32 \pm 0.58) \times 10^5$ psi. Tests on this granite in the laboratory yielded a rock substance modulus of 9×10^6 psi. Thus, in this case, the moduli of the rock mass in situ and of the rock substance differed by more than an order of magnitude.

(b) Strength

To interpret plate load tests in terms of strength parameters it is important to recognise the mode of failure; depending mostly on the rock type, a plate load test can produce (i) an instantaneous brittle failure or (ii) a plastic failure.

(i) Brittle Rocks

Failure occurs as soon as the condition for brittle failure at one point is satisfied. The first crack will propagate immediately and lead to an instantaneous fracturing of a limited zone beneath and around the plate accompanied by eventual rock bursting. It has been shown (3) that the average bearing pressure at failure, qf, is approximately related to the uniaxial compressive strength, Q_u, by:

$$qf = 3Q_u \qquad \text{........ (3)}$$

However, this equation is based on the assumption that the strength of the rock will be mobilised at the same time along the entire failure surface under the plate. For a test under a relatively rigid plate it is improbable that there will be sufficient yielding for this assumption to apply; the stress concentrations at the edge of the bearing plate are probably significant. Taking these stress concentrations into account the following approximate relationship was derived (3):

$$qf = 3\pi Qu \{y/B\}^{1/2}\{1-y/B\}^{1/2} \qquad \text{......... (4)}$$

where B is the width of the bearing area and y is the distance from the edge of the plate to the point under the plate at which fracture is initiated.

(ii) Yielding Rocks

Failure is manifested by a sudden yielding with circumferential cracks appearing close to the edge of the plate. The relationship between bearing pressure and cohesion has been expressed (3) by:

$$qf = 1.3cNc \qquad \text{........ (5)}$$

where c is the cohesion. Nc is the "bearing capacity factor" which is a function of the internal friction angle, ϕ, and is given by the simple, approximate expression:

$$Nc = 7 \tan^4(45+\phi/2) \qquad \text{........ (6)}$$

The angle of internal friction would usually be obtained from laboratory tests on samples which, if necessary, are recompacted.

In some cases it is difficult to distinguish which mode of failure is applicable as the fracture development is not always as distinct as ideally specified above. Equation 4 suggests that for brittle rocks under a rigid plate the bearing capacity will diminish with increasing width of the loading area whereas with yielding rocks Equation 5 suggests that the bearing capacity will be relatively unaffected by change in width of the bearing area. Hence conducting tests with various bearing widths and plotting the resultant bearing capacity against width might assist in identifying which interpretation for strength results should be used.

For example: Fig. 2c shows the results of a plate load test carried out on the wall of an iron ore mine. The rock exhibited considerable plastic irrecoverable strain and then the bearing pressure reached the value of 1500 psi, with a 5-in. dia. plate, the rock failed in the brittle manner. By plotting the bearing pressure at failure against plate size, as shown in Fig. 3, the dependence of bearing capacity on plate diameter is clearly seen, confirming that brittle failure was involved in these tests. As a comparison some results on yielding rocks (4) are also shown on this Figure; it is seen then the dependence on plate area is much less marked.

In general the interpretation of the results for brittle rocks is less satisfactory than those for yielding rocks owing to the inherent assumptions. Nevertheless, the plate load test can yield useful information, albeit crude, on the in-situ properties of the rock mass.

ROCK MASS MOVEMENT

Measurements of movement of a rock slope are one of the most sensitive indicators of stability. These measurements have the advantage that they are easily interpreted: an accelerating rate of movement indicates approaching instability; a constant or decreasing rate indicates present-day stability; and no movement represents the most stable condition. Surface movement can be measured using conventional surveying techniques, such as levelling, chaining, and triangulation. Movement at depth in the slope can be measured with borehole extensometers. Using both methods it should be possible to define the volume of rock which is moving.

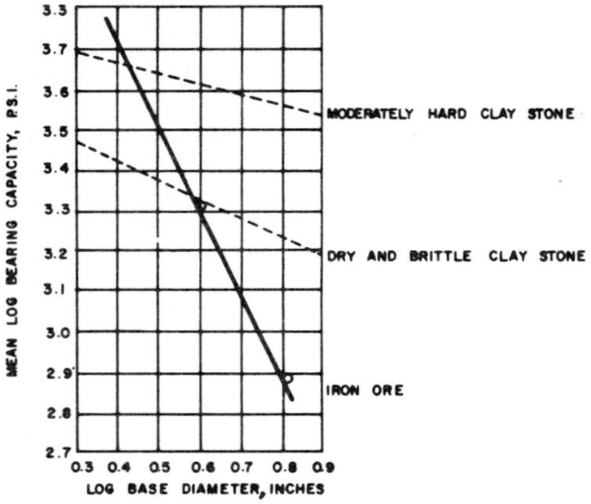

Fig. 3 —Dependency of bearing capacity on bearing-plate diameter.

Surveying Methods

For rock mechanics purposes surveying measurements usually have to be very precise, since change in distance is required rather than just distance itself. Only some of the modern instruments will be briefly mentioned since surveying methods are generally well-known and can be found in any standard textbook.

Levels which are self-levelling and incorporate a parallel-plate micrometer are convenient for rapidly measuring the subsidence of the slope crest. The micrometer attachment has a range of 0.01 ft. which covers the graduation on the stave. Measurements can be read to 0.0002 ft. (0.0024 in.). Manufacturers quote a mean-square error of 0.02 to 0.04 in. over a 3000-ft. double-run layout (5).

Three-dimensional movement of a point on a rock slope can be measured using a theodolite and a triangulation network. This method has been used to measure the deformation of power dams (6). A precise theodolite can measure to ± 0.5 seconds of arc, which for a 300-ft. sighting distance corresponds to approximately ± 0.01 in. The overall accuracy of the measurements depends on the triangulation layout, stability of stations, and precise mounting, sighting, and reading of theodolite.

Borehole Extensometers

One of the simplest methods of measuring rock mass movement is with a borehole extensometer. Basically the extensometer measures the relative axial movement between a point in a borehole and the borehole collar. A wire, or solid bolt, spans these two points and is used as a standard reference length. Drilling a borehole is one of the major costs, consequently single bolts are usually not installed in boreholes greater than 20 ft. Up to four wires (anchored at different depths) have been installed in boreholes 300 ft. long.

In selecting an extensometer for measuring rock mass movement, it is necessary to define the type of movement which is to be measured, since this determines the required accuracy of measurement. Rock movements produced by changes in stress result in relatively small displacements, whereas those produced by fracturing, sliding along cracks, fissures and joints, etc. are relatively large. Obviously the accuracy of the measurements needs to be greater in the first case than in the second. Movement of fractured rock can adequately be measured within an accuracy of 0.050 in., and movement due to change in stress should be measured within an accuracy of at least 0.010 in. These accuracies can be easily achieved for short measurement lengths (up to 50 ft.) with almost any of the available extensometers. In this respect, deformation due to change in stress is rarely measured in boreholes over 100 ft. long.

Although a borehole extensometer measuring system is basically simple, there are a number of errors, such as those produced by friction and temperature changes, which can affect measurements. If these errors are significant and not recognized, then the measurements can be misleading.

It has been shown (7) that friction within the extensometer can produce excessive errors. Friction between the wires, anchors, and boreholes can be kept within acceptable limits by permanently separating the wires and using anti-friction bearings where they pass through the anchors or spacers. Installing too many wires in one borehole, and especially if they pass through a central hole in the anchors, can lead to excessive interference. Wire diameter should be as large as practical (say 0.049-in. dia. or more) since it reduces the effect of friction.

Temperature changes affect both the length of the borehole and wire, and can produce large errors. Temperature affects boreholes drilled from the ground surface to a depth of approximately 25 ft., below which the temperature remains almost constant. In many cases, temperature variations can be corrected by installing a wire at a 25-ft. depth and subtracting the movement on this wire from the movements recorded by other wires. To obtain the movement of the borehole, including that produced by temperature, two wires with different coefficients of linear expansion can be connected to the same anchor at 25 ft.

There are various types of extensometers. One of the simplest is a dial gauge which measures the distance between a plate anchored at the collar of the borehole and the end of a solid bolt. For wire extensometers, either a constant tension system can be used or a fixed wire producing varying tension as the anchor moves. In addition, electrical readout systems can be used.

If possible, preference should be given to constant-tension extensometers since they are inherently more accurate than variable-tension extensometers. The latter type are not only affected by friction and temperature change but also by change in the physical properties and sag of the wire. The apparent elastic modulus of a wire changes with applied tension due to the straightening out of the natural curvature of the wire. In some cases at a tension of 5 lb, the elastic modulus is only a third of that at 15 lb. Consequently, the corrections due to change in wire length with change in tension are not straightforward, and the instruments require extensive calibration with different wire lengths prior to being used in the field.

With constant-tension extensometers, wire tension is adjusted to a constant value every time a reading is taken, and no correction factors are required. This type of extensometer has an additional advantage in that readings can be taken at two tensions, thus measuring the stretch of the wire. The stretch should be the same every time readings are taken and should correspond to the length of the wire. A reduction in stretch indicates that the wire is obstructed, and the place where obstruction occurs can be calculated. Consequently, reading at two tensions indicates how good the installation has been and the accuracy of the measurements; it prevents erroneous interpretation of the results.

Figs. 4 and 5 show two extensometers currently used by the Mining Research Centre.

Fig. 5 —Constant torque electrical readout extensometer.

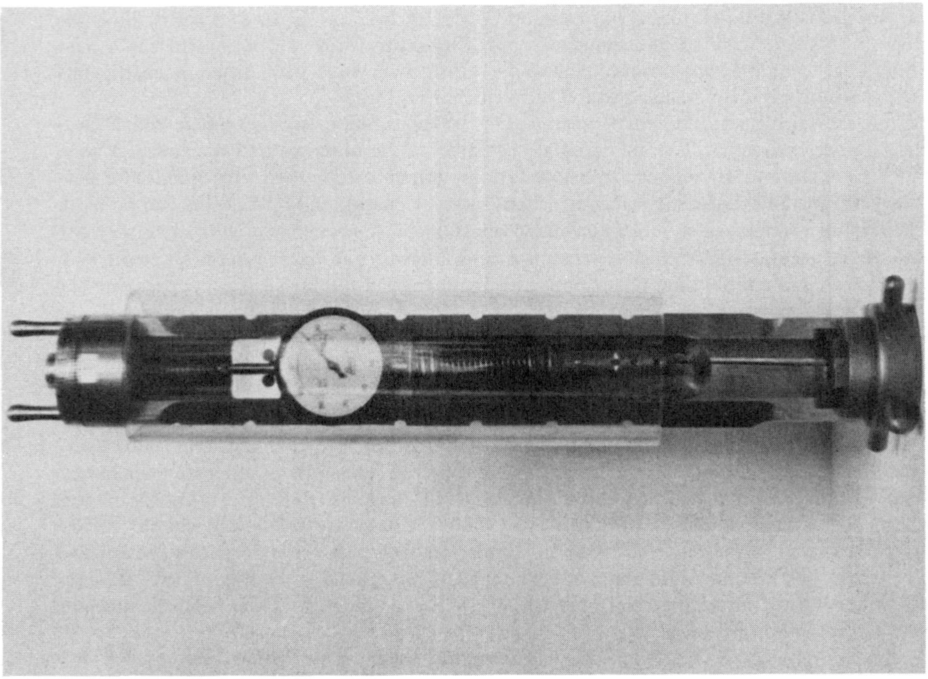

Fig. 4 —Newcastle MK-II manual readout extensometer.

The Newcastle Mk II extensometer is a manual constant tension type instrument which may be moved from borehole to borehole. Tension is exerted through an extension spring, and a dial gauge measures spring deflection and hence tension. A scale on the side of the instrument (graduated in inches and tenths) and the handle (graduated in thousands of an inch) are used to measure the position of the wire.

Most current designs for remote readout extensometers are variable-tension instruments. Tension is usually supplied by helical springs or cantilevers; rotary potentiometers, linear potentiometers, LVDT's, vibrating wires and strain gauges have been used for sensing the tension change and relating it to movement. Recently a new type of constant-force spring (NEG'ATOR — Hunter Spring Co.) has become available. Fig. 5 shows an electrical readout constant-tension extensometer designed at the Mining Research Centre. Constant-torque springs exert a force to four pulleys, to which flexible wire cables are attached. As the wire moves the pulley rotates and this rotation is measured with a rotary potentiometer. The circumference of the pulley is 2 in. and the readout unit has 1000 divs, hence the instrument reads to 0.002 in. After the wiper on the potentiometer passes through a small gap in the resistance coil, the readings start off at zero again. Hence the range is almost unlimited, provided a movement of more than 2 in. does not occur at any one time. Also, the instrument can be connected to give a continuous readout. Laboratory tests show that the tension exerted on the wire is 16 lb $\pm$0.25 lb over a 5 in. range. Field trials are being conducted with this instrument.

If the measuring location is inaccessible or conditions are hazardous a remote readout constant or variable tension extensometer is required. For readily accessible and safe locations a manual readout constant tension extensometer has some advantages, namely (a) the readings required for any number of boreholes, hence it is cheaper; (b) the readings are usually more accurate than with electrical extensometers; (c) by taking readings at two tensions, the condition of the wires in the boreholes can be evaluated.

Shear-strip Indicator

An entirely different method of detecting rock mass movement is with a shear-strip indicator. The shear strip is comprised of a number of separate electrical resistors mounted 6 in. apart on a bakelite strip and wired in parallel. A cable is attached to both ends of the strip, and the whole assembly is grouted into a borehole. A readout unit indicates directly the number of resistances in the circuit. Once grouted into position a differential shear movement of approximately 0.1 in. is sufficient to shear the strip, breaking the circuit. Taking readings from both ends of the circuit enable the depth at which shearing has occurred to be calculated. Hence the width of the shear zone can be determined, and a distinction can be made between a shear plane and a shear zone.

Surface monitoring during failure

Previous sections describe borehole instruments used for obtaining early warning of incipient failure. If a slope is failing then it is even more important to monitor the movement of the failing mass. In such cases the movement is generally much larger and thus the accuracy required for measurement need not be as high. Very useful information on the displacement and acceleration of a failing rock mass can be obtained from a few simple, crude, devices which cost only a few dollars. For example, if stakes are placed on either side of a tension crack in the crest of a slope their relative movement can be measured by attaching a wire to the stake in the moving mass, running it over a pulley on the stake in the stable ground and weighting the end of the wire. The position of the weight relative to the stake can be measured with even a ruler to approximately $\pm 1/10$ in. giving a measure of the opening of the tension crack. Many other equally simple and crude devices have been similarly used, their design usually depending only on the materials at hand and on the ingenuity of the engineer.

STRESS DETERMINATION

Stress determination in rocks can be divided into two categories; firstly, the determination of the absolute stress in rock by stress relieving a portion of rock and secondly, the monitoring of stress changes due to excavation. Since at present no single technique is suited to both measurement categories, they will be considered as separate problems.

Determination of Abolute Stresses in Rock

Two methods for the determination of abolute stress in rock are currently in use in Canada. The first measures the change in diameter of a borehole when it is overcored whilst the second measures the change in strain on the flattened end of a borehole due to overcoring. To calculate the stress from these measurements, both methods assume elastic behaviour of the rock, and hence Young's modulus, E, and Poisson's ratio, υ, must be determined from standard laboratory sample tests.

(a) Borehole Deformation Measurements

If U is the change of diameter of a hole of diameter d and P, Q and θ are the secondary principal stresses and directions in the plane normal to the axis of the hole, then:

$$U = \frac{d}{E} \{(P+Q)-2(P-Q)\text{Cos } 2\theta - \upsilon\sigma_z\} \qquad \ldots\ldots\ldots (7)$$

where σ_z is the axial stress along the borehole. U_1, U_2 and U_3, across three diameters of the hole oriented at angles θ_1, θ_2 and θ_3, are measured when the hole is overcored. Substituting these results into Equation 7 enables the secondary principal stresses, P, Q and direction, θ, to be calculated only if some assumption is made for the axial stress, σ_z. Sometimes it is valid to assume $\sigma_z = 0$; this is probably true in thin pillars and also in a borehole normal to and close to a free surface. Sometimes the direction θ may be assumed and P, Q and σ_z calculated. If no assumptions can be allowed then at least one more independent diameter change must be measured in another hole oriented in a different direction.

The above analysis refers to the determination of the secondary principal stresses in a plane. To determine the complete three-dimensional stress field it is necessary to make a minimum of six independent diameter measurements in a minimum of three different obliquely oriented boreholes (8).

One deformation meter, developed by the U.S. Bureau of Mines (9), uses strain-gauged cantilever beams to sense the deformation across the diameter of an EX borehole. The output of the strain gauges are calibrated directly in terms of deformation. This meter is installed at the required depth and then overcored with a 6 in. coring bit, as illustrated in Fig. 6a. Readings are taken continuously during overcoring, the difference between the initial and final readings giving deformation changes. Fig. 6b shows a typical curve of deformation change versus bit penetration during overcoring. Care must be taken to ensure that overcoring is continued sufficiently past the meter so that the true deformation change, U, shown in Fig. 6b is obtained; premature stopping of coring at, say, position 2 in this figure would yield a false reading, Ua. Likewise the meter must be pushed sufficiently far into the hole at the start of measurements; if, say, a start was made at position 1 in the figure then a false reading, Ub, could be obtained. (Note: the graph in this figure is not necessarily always this shape; nonetheless these rules on meter insertion and complete overcoring still apply.) Fig. 6c shows a set of overcoring deformation data.

(b) Strain Measurements on the Ends of Boreholes

If a 45° strain-gauge rosette is glued to the flattened end of a borehole and e_x, e_{xy} and e_y are the strain changes measured by overcoring the gauge, then the stress components σ_x, σ_y, σ_z and τ_{xy} are given by:

$$e_x = A\sigma_x + B\sigma_y + C\sigma_z$$

$$e_{xy} = \tfrac{1}{2}(A+B)\,(\sigma_x - \sigma_y) + C\sigma_z + (A-B)\,\tau_{xy} \qquad \ldots\ldots\ldots (8)$$

$$e_y = A\sigma_y + B\sigma_x + C\sigma_z$$

and the secondary principal stresses in the xy plane are given by:

$$P, Q = \frac{\sigma_x + \sigma_y}{2} \pm \sqrt{4\tau_{xy}^2 - (\sigma_x - \sigma_y)^2}$$

and $$\tan 2\theta = \frac{2\tau_{xy}}{(\sigma_x - \sigma_y)} \quad \ldots\ldots\ldots (9)$$

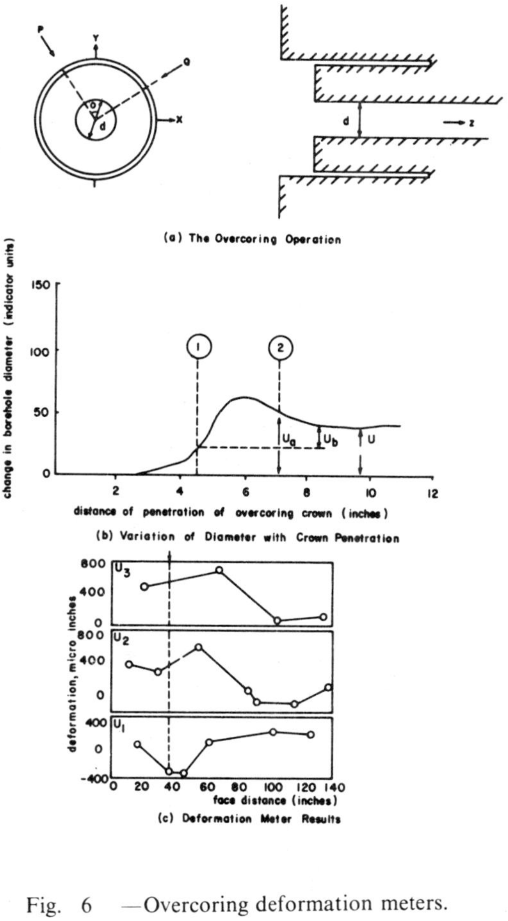

Fig. 6 —Overcoring deformation meters.

As with the deformation meter, P, Q and θ can only be calculated from the measured strains if some assumption is made for the axial stress, σ_z. If such an assumption is invalid then measurements must be made on the end of another differently oriented borehole in at least one more independent direction in order to calculate P, Q and θ. To determine the three-dimensional stress field, measurements must be made in at least six independent directions in a minimum of three oblique boreholes.

The factors A, B and C appearing in Equation 8 are given by:

$$A = (a - b\upsilon)/E \; ; \; B = (b - a\upsilon)/E \; ; \; C = c(1-\upsilon)/E \qquad \ldots\ldots\ldots (10)$$

where a, b and c are the stress concentration factors on the end of a borehole and are given (10) by:

$$a = 1.366 + 0.025\upsilon + 0.502\upsilon^2$$

$$b = -0.125 + 0.154\upsilon + 0.390\upsilon^2 \qquad \ldots\ldots\ldots (11)$$

$$c = -0.520 - 1{,}331\upsilon + 0.886\upsilon^2$$

One suitable instrument for this type of measurement is the CSIR strain cell (11). As illustrated in Fig. 7c, this cell consists of a foil strain-gauge rosette which is encapsulated with a plug. The cell is bonded with strain-gauge cement to the flattened end of a BX borehole and is then overcored with a BX coring bit, as illustrated in Fig. 7a. Readings of the three strain gauges before and after overcoring gave the values, ex, exy and ey. The installation tool, Fig. 7b, contains a mercury orienting switch and is spring-loaded to ensure application of the

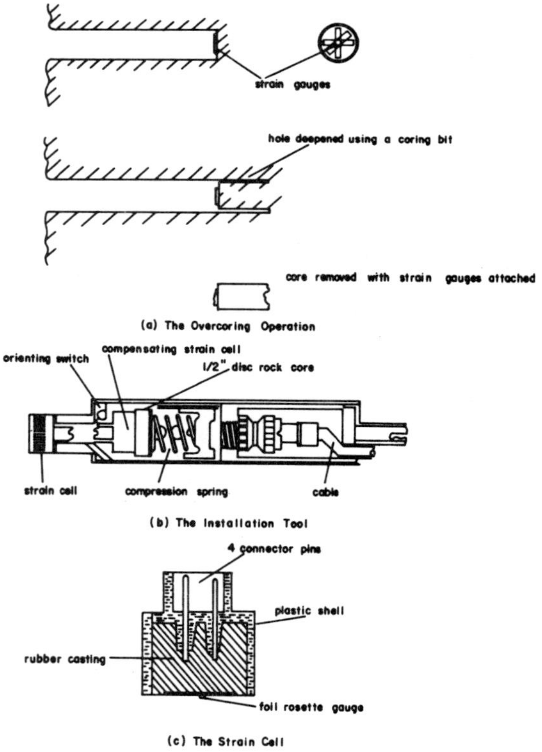

Fig. 7 —Overcoring a strain cell.

correct force (5-10 lb) during curing of the bonding cement. The tool also contains a dummy strain cell mounted on a piece of core to act as a temperature compensator in the strain-gauge bridge network.

The initial strain reading is taken immediately prior to overcoring and the final reading is taken by replugging the cell into the tool after the readings may drift for 1 — 1½ hrs. as the rock temperature adjusts to that of the installation tool. The final strain reading is taken when this drift ceases. A continuous drift of readings for much longer than this time usually indicates that the cell is mounted on a piece of cracked core, in which case the readings should be discarded and the operation repeated.

Since this method relies entirely on an accurate measure of the relieved strain in the rock, it is essential that the bond of the strain-gauge rosette is checked to ensure that valid strain readings have been obtained. This is done by cutting off a 1 in. thick disc of core, with the cell still attached, from the overcored specimen and loading this disc diametrically in the direction of each gauge in turn. An inadequately bonded gauge will show one or more of the following symptoms: (a) a very poor sensitivity; (b) a sudden breakdown in response to load; (c) a large hysterisis loop (which is not a rock property) or (d) a continuous zero shift with each load cycle.

The use of four-element strain rosettes, rather than three-element rosettes, can assist considerably in the assessment of the reliability of the results (12). Methods developed for statistically handling the overcoring data also help considerably in eliminating erroneous data and in assessing the accuracy of the results (12).

(c) Comparison of the Methods

The CSIR strain cell has the advantage that the overcoring size is relatively small and hence drilling costs are reduced. The cell itself is inexpensive and is thus expendable. The method may be used in deep holes (300 ft. or even more). However, it suffers from the disadvantage that it is unsuitable for use in wet conditions since strain-gauge cements do not bond well to wet rock. Consequently, this method is not generally used in holes dipping below the horizontal.

Deformation meters are much better suited to wet conditions and can be used in down-dipping holes (although in long holes dipping steeper than 45°, problems may be experienced due to drill cuttings binding between the meter and the hole). Coring the 6 in. hole results in relatively high drilling costs; consequently, this method has not been used in holes much deeper than 100 ft. The deformation meters are expensive and not expendable; however, with care they may be used many times.

Both methods suffer from the fact that they cannot be used in deep vertical holes; this is a particular disadvantage for open pit problems where stress measurements in vertical holes around the periphery of the pit, away from the influence of the excavation, would be particularly desirable to define the stress field acting on the pit.

A word of warning should be added here. Both these methods are expensive, and although their principles are simple, the techniques for field measurement and handling of data demand considerable "know how". The reader is thus advised to seek professional advice before undertaking this type of study.

Monitoring of Stress Changes in Rocks

A relatively cheap device for monitoring stress changes in rocks is the photoelastic stress meter (13, 14). This meter consists of a 1½ in. dia. glass cylinder with a central circular hole of ¼ in. dia. At one end a light source and filters producing circularly polarised light are attached. The cylinder and light source are grouted into a borehole and are viewed with a telescope through a ¼-wave plate and analyser. Stress changes in the rock induce a distinct coloured fringe pattern in the meter. The fringe order at any point in the meter can be determined by counting the fringes, as illustrated in Fig. 8; fractional fringes can be determined by the Tardy compensation method.

To interpret the fringes in terms of the magnitude and direction of the secondary principal stresses the following procedure is used. The direction of the secondary principal stresses coincides with the axes of symmetry of the fringe pattern. The ratio of the principal stresses, $Q/P = \zeta$, is then determined. This is done by looking for the two black zero-points (isotropic points) which occur on the major principal stress axis (Figs. 8b, c and d). If these isotropic points do not occur then either the stress is almost uniaxial ($Q = 0$) or the stress is hydrostatic ($P = Q$) in which case the fringes are circular. Assuming that these isotropic points do occur, then the ratio ζ can be determined by measuring the diameter Do between them and comparing it with the diameter Dh of the central circular hole. Fig. 9a gives the values of ζ for various Do/Dh values, e.g., if Do/Dh = 3.0, $\zeta = Q/P = 0.75$.

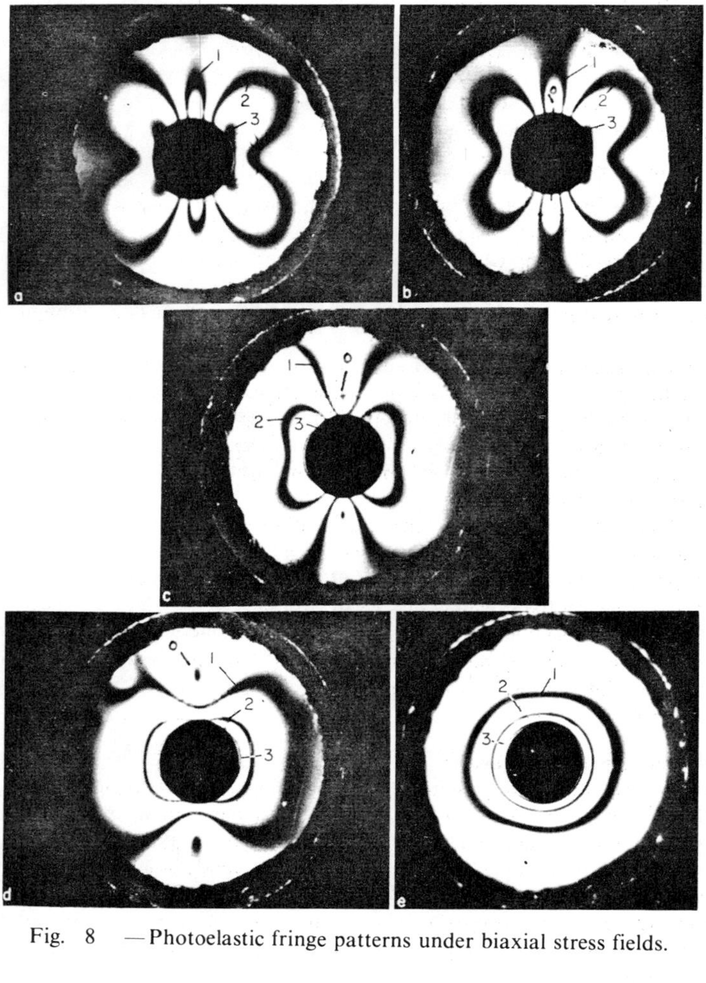

Fig. 8 —Photoelastic fringe patterns under biaxial stress fields.

It now remains to determine P. This is done by measuring the fringe order at particular points in the meter. Since there are optimum points at which these measurements should be made for best accuracy (14) the following rules should be followed:

(a) if $\zeta \leqslant 0.33$. The fringe order, n_1, should be measured at a point on the $\theta = 45°$ axis at a radius of r = 0.20 in. P is given by:

$$P = \frac{n_1 f}{t}\left\{\frac{1}{\alpha(1-\zeta^2)+\beta(1+\zeta)^2}\right\}^{1/2} \qquad \text{........ (12)}$$

where α and β depend on the rock modulus and are given by Fig. 9b. f = 1030 psi/fringe which is the fringe constant for glass and t is the thickness of the glass meter.

(b) if $\zeta \geqslant 0.33$. The fringe order, n_2, should be measured at a point on the $\theta = 0°$ axis at a radius r = 0.175 in. P is then given by:

$$P = \frac{n_2 f}{t}\left\{\frac{1}{(A-B\zeta)}\right\} \qquad \text{........ (13)}$$

where A and B depend on the rock modulus and are also given in Fig. 9b.

This instrument has been successfully used in holes of 10-20 ft. deep. For deeper holes, installation techniques and viewing methods require further development.

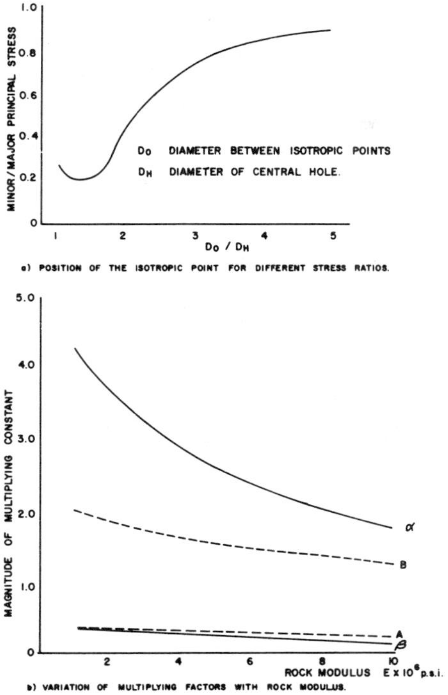

Fig. 9 —Interpretation of photoelastic fring patterns.

GROUNDWATER MEASUREMENTS

Groundwater can affect open pit operations. If the prevailing groundwater level in the surrounding country is higher than the pit bottom, seepage will occur under the wall slopes causing additional pressures to encourage sliding. Where such seepage breaks out on the surface of a slope, erosion in weaker rocks can not only affect the stability of surface layers but also affect operations by the addition of mud to the roads. The measures required to eliminate these unsatisfactory effects can be a significant factor in the economics of the operation.

Groundwater Elevation

The primary groundwater measurement is the determination of the elevation of the free water surface. Whereas, if water is a problem, it will be valuable to measure the groundwater elevation wherever possible, e.g., in exploration holes, blastholes and holes for structural drilling, it may be necessary to establish observation wells (sometimes called piezometers, which simply means a meter for measuring the piezometric head of water, i.e., position head plus pressure head) at certain locations. In some rocks and in particular locations where surface runoff is not likely to enter the hole, it is satisfactory to drill a vertical hole and allow it to stand without casing. The size of the hole depends substantially on the length of the hole and on any difficulties that are experienced in maintaining it vertical, e.g., an AX hole could be used possibly to a depth of 50 ft. in favourable circumstances; a 6 in. hole would be required for a depth of a few hundred feet. In other cases, where the walls of the hole might be unstable or where it is desired to isolate one layer of rock or where surface runoff is to be excluded from the hole, casing will be required as shown in Fig. 10b.

Where the rock is unstable at the elevation of inflow, it will be necessary to use the type of installation shown in Fig. 10c. A porous stone, which is commercially available, is connected to flexible tubing of approximately ½ in. dia., which can be of either saran, nylon, or metal. The porous stone is lowered into the hole and centered with the aid of a spider. Sand or gravel, which is tamped lightly with an annular ring on a rod, is poured into the hole for the depth of the rock zone to be tapped by the observation well. On top of the sand, a seal is created with either clay, bentonite or grout. The clay and bentonite, when used, are placed by dropping plastic balls and tamping. With this type of installation several porous stones, or observation points, can be installed in the same hole.

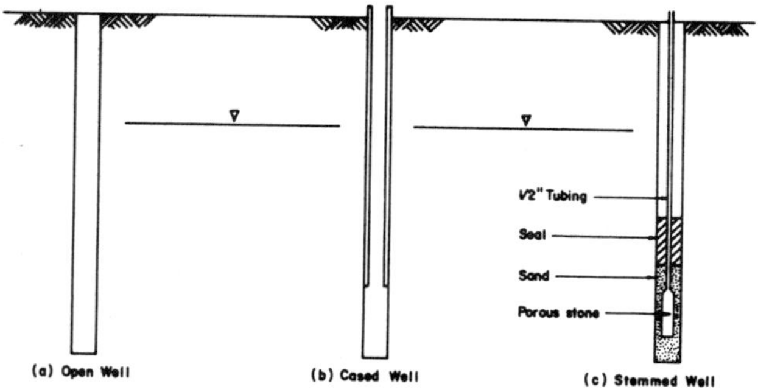

Fig. 10 —Observation wells for groundwater measurements.

The simplest and most direct way of measuring the elevation of the water in the observation well is by using a hand-held sounder, which is commercially available, such as shown in Fig. 11a. The cable, consisting of two wires, is calibrated in feet or metres, and the probe consists of a weight, normally lead, which holds the bare ends of the two wires slightly apart. On contact with the water the galvanometer located on the side of the reel shows that the battery circuit has been closed (a drop of water in the side of the tubing may sometimes cause an erroneous reading).

For open wells automatic recorders such as shown in Fig. 11b can be used. Here a counterbalanced float controls the position of the drivewheel so that the pen driven by a clock records the water level with respect to time. The holes must be at least 5 in. in dia.; however, the continuous record can be very valuable.

In soil mechanics, for stability reasons it is usually of more interest to know the piezometric pressure of the water in relatively impermeable clay strata. Owing to the great time lag that would be required to register change in pressure with an observation well such as shown in Fig. 10b, instruments have been developed that will sense the pressure without the requirement for flow. However, in rock mechanics this requirement seldom occurs.

Other types of sensing units are available. For example, air pressure can be used to drive the water down to the bottom of tubing inserted in the well, and the amount of pressure required to balance the head of water read from a conventional gauge. In a deep well using bottled compressed air, there can be considerable saving in time required for each reading compared to using the hand-sounding apparatus of Fig. 11a.

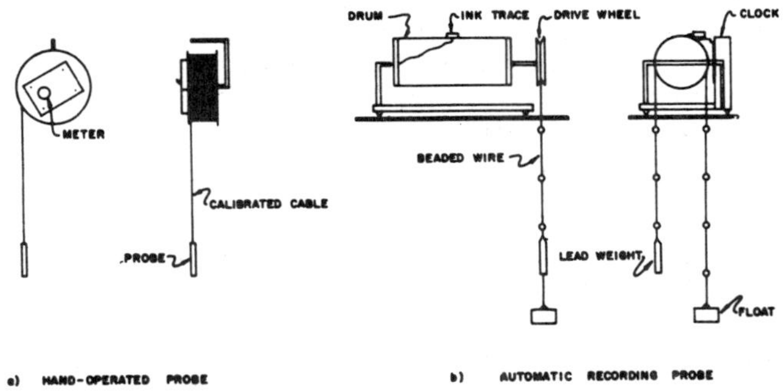

Fig. 11 —Sensing of water elevations in observation wells.

In-Situ Permeability and Drawdown

For determining feasibility and design of water control systems, it is helpful to have some measure of the permeability of the ground carrying the water (15). The simplest and cheapest test that can be conducted is a falling head test. It is assumed for this test that the ground is equally permeable in all directions and consequently the water flows out from the bottom of the hole into a spherical volume. The borehole must be cased to within a distance of the bottom of the hole equal to its diameter. The casing is filled with water, which is allowed to fall to the groundwater level, several times; depths to the water surface being observed after the start of the test. The coefficient of permeability, k, is obtained from the following equation:

$$k = 2.3r \log (H_1/H_2) / (4(t_2 - t_1) \qquad \text{........ (14)}$$

where r is the radius of the hole, H_1 and H_2 are the heads over the groundwater level at times t_1 and t_2 respectively, the units usually being in cm and sec. Alternatively, with a method of measuring flow of water, a test can be run such that the flow, Q, just required to maintain the head in the casing is determined. The coefficient of permeability is then obtained from the following equation:

$$k = Q/(4\pi Hr) \qquad \ldots\ldots\ldots (15)$$

where H is the head over the water table. Although these tests are simple and cheap, their value is questionable as the volume of rock being tested is small and the possibility exists, without special precautions being taken, for the water to flow up between the casing and the sides of the hole.

Similar types of tests have been used with water being pumped under pressure into the rock, usually into preselected zones that are isolated by packers. If relatively high pressures are used, expansion of the hole and the joints carrying water can be significantly affected, which results in the coefficient of permeability determined from the test being too large.

For designing a drawdown system, the best test is a drawdown pumping test. This test consists of a pump well together with several observation wells. Whereas it is preferable to have the observation wells along a line connecting the pump well as shown in Fig. 12a, this is not always practical, and from a theoretical point of view, assuming a uniform water-bearing layer, it should not be essential. On the other hand, if the water-bearing layer is not uniform in all horizontal directions, it may be very important to have several lines of observation wells to obtain information on the differences in drawdown with respect to direction. The simplest procedure is to pump at a known rate until the water levels in the observation wells become static. Assuming that the water is coming from a layer of ground of uniform thickness, that the well has fully penetrated this layer, and that the drawdown at the pump well is small with respect to the thickness of the layer (all required for radial flow), the coefficient of permeability can be determined from the following equation:

$$k = \frac{0.366Q \log(b/a)}{zh_{ba}} \qquad \ldots\ldots\ldots (16)$$

where b and a are radial distances to two observations wells, H_b and H_a are the heads of water in the observation wells, h_{ba} is the difference of drawdown between these two wells equal to (H_b-H_a), and z is the thickness of the water-bearing layer. If the upper surface of the water-bearing ground is not restricted and the drawdown of the pump well is not small, a corrected differential head should be used in the above equation as determined from the following formula:

$$h_{ba}^{1} = h_{ba} - h_{ba}^{2}/(2z) \qquad \ldots\ldots\ldots (17)$$

Whereas the thickness of the water-bearing layer should be known, in fact, particularly on mining properties, rock formations are often not that tidy. In these cases, pumping tests should still be conducted but they should be designed with a more empirical orientation, i.e., they should be located as close as possible to the zone where the system would be operating. The directional properties of flow and drawdown, as mentioned above, also become more important. Not knowing z, the above formula can be modified as follows:

$$h_{ba} = cQ \log (b/a) \qquad \ldots\ldots\ldots (18)$$

where c contains the various constants in the above equation, which may not be constant for conditions other than radial flow. Nevertheless, such pumping tests would still provide a basis for designing a system for an open pit even though the degree of accuracy of the prediction would not be high.

As an example, a test was conducted in a water-bearing layer, z, of 80 ft. using a 24 in. pump well. The pumping rate was 175 gpm (0.468 cfs). The elevations in the observation wells at equilibrium were as shown in Fig. 12b. The slope of this curve indicates that a drawdown, H_{ba}, of 1.8 ft. occurs between radial distances of 10 ft. and 100 ft. The coefficient of permeability is then computed from Equation 16 as follows:

$$k = \frac{0.366\ \beta\ 0.468}{80\ \beta\ 1.8.} = 0.0013\,\text{ft/sec.}$$

By testing the correction that should be applied for the approximate 6 ft. of drawdown at the pump well, it is found that only a 4% error is introduced for this reason, which can be ignored.

Based on this test, the drawdown at the pump well can be computed for a pumping rate of 1000 gpm (2.67 cfs). The radius of influence of the well was determined from the test to be 2000 ft. The drawdown at the pump well is then computed as follows:

$$h_{ba} = \frac{0.366\ \beta\ 2.67 \log 2000/1}{80\ \beta\ 0.0013} = 29.0 \text{ ft.}$$

In this case, the drawdown is not small with respect to the thickness of the water-bearing layer; consequently, the actual drawdown is determined from the Equation 17 above as follows:

$$29.0 = h_{ba} - h_{ba}^2/160 \qquad h_{ba} = 38 \text{ ft.}$$

As the drawdown for radial flow is proportional to the pumping rate, the curve in Fig. 12b for 1000 gpm can be used for approximate extrapolation to other pumping rates for the determination of the drawdown at various distances from the pump well.

Whereas in most cases a pumping test is preferable to other tests, if the ground is very erratic many cheap tests might be better than a few expensive tests. In other words, many falling head or pressure pumping tests could provide a better appraisal of the variability of the permeability and thereby be a better basis for determining the feasibility of a drawdown system.

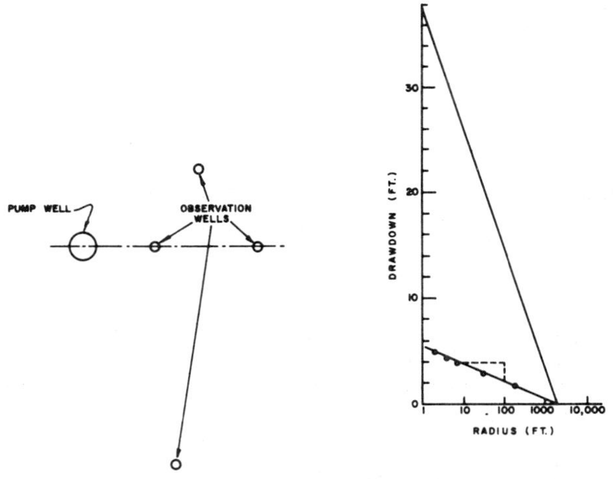

a) TYPICAL WELL LAYOUT FOR DRAWDOWN TEST.

b) EQUILIBRIUM WATER ELEVATIONS IN OBSERVATION WELLS DURING DRAWDOWN TESTS.

Fig. 12 —Drawdown pumping test for in-situ permeability.

BLASTING MEASUREMENTS

The walls of an open pit can be affected by the blasting practices being used. Measurements of various aspects of the blasting operation can provide useful information for developing procedures that will optimize economy of both the rock breakage and ground control objectives. Such measurements, unfortunately, are not as easily performed as those of ground movement or water levels. However, with some guidance and good equipment, there is no reason why the technical staff of most mines could not make use of such techniques.

Detonation Velocity

The proper functioning of the explosive in the blasthole is sometimes questioned. One of the easiest ways of determining whether the explosive is performing as expected is to measure the detonation velocity. Fig. 13a is a photograph of an explosive probe that has been developed for determining the detonation velocities (16).

The explosive probe consists of two circuits. One circuit, which acts as a trigger, consists of a pair of insulated wires running to the end of the unit where the ends of the wires form an ion gap that is shorted by the plasma of the detonation front causing the oscilliscope trace to be started. The second circuit consists of a chain of resistors that are sequentially shorted to ground when they are enveloped by the detonation front. This circuit is connected to a constant current supply that produces a step wave form as a result of the shorting action. The sequence of steps produced on the oscilloscope is photographed, which provides the record for determining the detonation velocity.

The cylindrical body of the probe consists of ¼ in. hard brass tubing with a wall thickness of 0.010 in. Belden cable is used to provide two pairs of shielded conductors with a ground wire. The trigger circuit has a 220-k ohm resistor between the ends of the wires, which provides a means of checking the continuity of the circuit. The current circuit consists of one conductor connected to the chain of ½-watt carbon resistors and the other conductor together with the uninsulated ground wire of the cable, which are soldered to the brass sleeve. The end of the resistor chain is soldered to the ground side of the 220-k ohm resistor at the end of the trigger circuit.

The probe is made waterproof by soldering at the end and by crimping the collar of the brass tubing and filling the void with an epoxy resin compound.

Fig. 13b shows a typical voltage stepwave form. In this case, the step length varies, which indicates that the detonation velocity of the explosion is varying, possibly due to unevenness in packing the 50% straight dynamite.

Rock Acceleration

The acceleration produced in the adjacent rock by a blast can be measured. The amount of stress that the wallrock is subjected to, can be of interest, e.g., for examining the effectiveness of special perimeter blasting patterns. Also, acceleration records can be used to detect misfiring or incorrect delay intervals.

Accelerometers are available commercially that provide a shock capacity of 10,000g or more. The voltage sensitivity is of the order 10 mV/g. The frequency response should be in the range between 5 Hz and 10 kHz, and the calibration information supplied by the manufacturers should be augmented by shock calibration.

A convenient mounting assembly for accelerometers is shown in Fig. 14a. A hydraulic hand pump, operated on the surface, is used to wedge the mount in the hole with a pressure of 2000 psi. The sleeves and wedges can be made of aluminum to provide an approximate acoustical match between the accelerometers and most hard rock. The mounts are oriented so that the sensitive direction of the accelerometer is towards the direction of the blasthole of greatest interest.

The output from each accelerometer is put through a cathode follower, with the short-

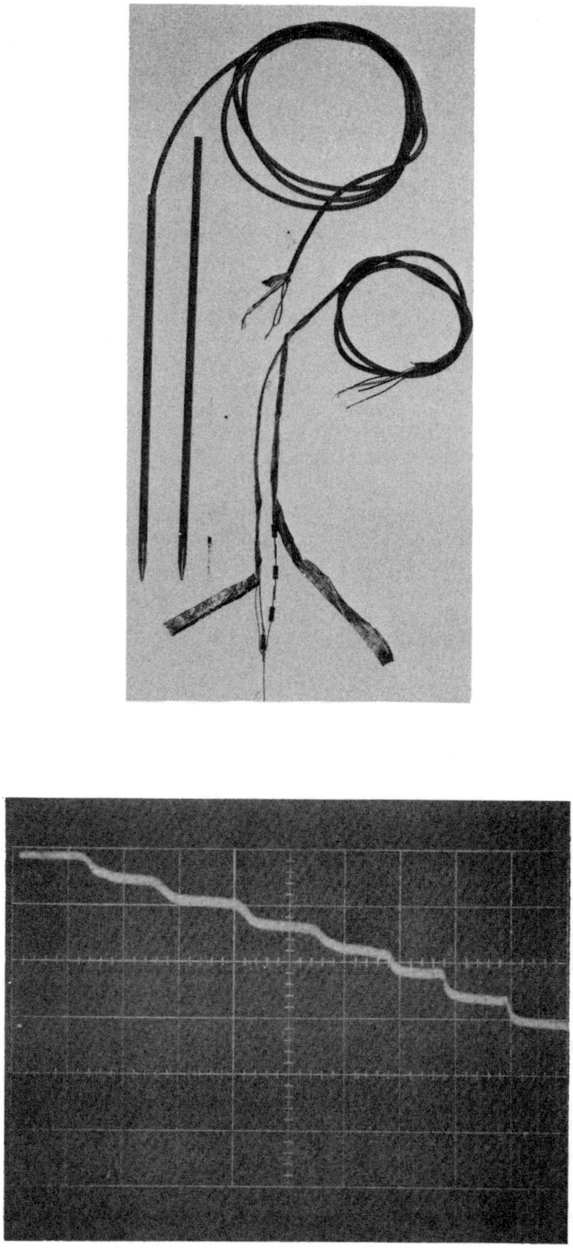

Fig. 13 —Detonation velocity measurements.

est possible cable intervening, so that the signal can be transported for up to a 1000 ft. to the recorder. The recording apparatus, normally consisting of an FM tape recording unit, can be controlled by a "one cycle and stop" timer which is initiated by the closing of a push-button contact. The tape recorder requires a runup time of four seconds before detonation. The outputs from the accelerometers are then recorded for an additional four seconds after which a reference signal from an oscillator is recorded.

On playback the outputs of the accelerometers can be put through an oscilloscope and photographed for a visual record. In addition, using an operational amplifier it is possible to integrate the acceleration wave to obtain the particle velocity wave. Fig. 14b shows an example of such an integration. Knowing the peak particle velocity, v, a measure of the peak radial stress, σ from the blast can be determined by using the following equation:

$$\sigma = \rho \, v \, Cp \qquad \text{......... (19)}$$

where ρ is the mass density of the rock and Cp is the velocity of the P-wave in the rock. Alternatively, the maximum strain can be calculated as follows:

$$\epsilon = v/Cp \qquad \text{......... (20)}$$

A second integration, shown in Fig. 14c, similarly produces a displacement record. Depending on the purposes of the measurements, either stress, strain or displacement can then be used for comparing the blasting effects.

In a bench in an open pit the fragmentation and the shape of the muckpile indicated that something had gone wrong with the blast. The accelerometer record indicated the cause of the unsatisfactory results. Fig. 15 shows the sequence of 15-ms (millisecond) rated delays (actually closer to 20 ms) between holes, and the arrow shows that one hole did not fire.

ANTICIPATED FUTURE DEVELOPMENTS

Electro-Optical Surveying

As mentioned previously (section 4), conventional surveying instruments can be used to measure slope movement. Recently electro-optical distance-measuring instruments have become commercially available which measure distances over several miles to sufficient accuracy for rock mechanics purposes (17). Basically a modulated beam of light is transmitted from the instrument to a reflector on the rock slope, which reflects it back to the instrument. The transmitted and received beams are compared, and the phase shift, which is proportional to the distance, is measured.

A typical measuring network for an open pit is shown in Fig. 16. Three instrument stations are located near the crest of the pit. Since these stations are likely to move over a period of time, they are tied into a system of three stable reference points. The distance is measured optically from each station to the reference points. Knowing the coordinates of the reference points, the three-dimensional coordinates of the measuring stations can be calculated (18). Then using the instrument stations as reference points the distances to reflector stations on the pit slope are measured and the coordinates calculated.

Since twelve distance measurements are required to determine the coordinates of one point on the pit slope, the geometrical configuration of the network is important with regard to overall accuracy. In particular, if there is insufficient vertical elevation between the reference and instrument stations, the vertical angle would have to be measured with a theodolite.

Complete Stress Determination from Measurements in a Single Hole

The methods for absolute stress determination, described in Section 5, require measurements in three oblique boreholes if the three-dimensional stress field is to be determined. How-

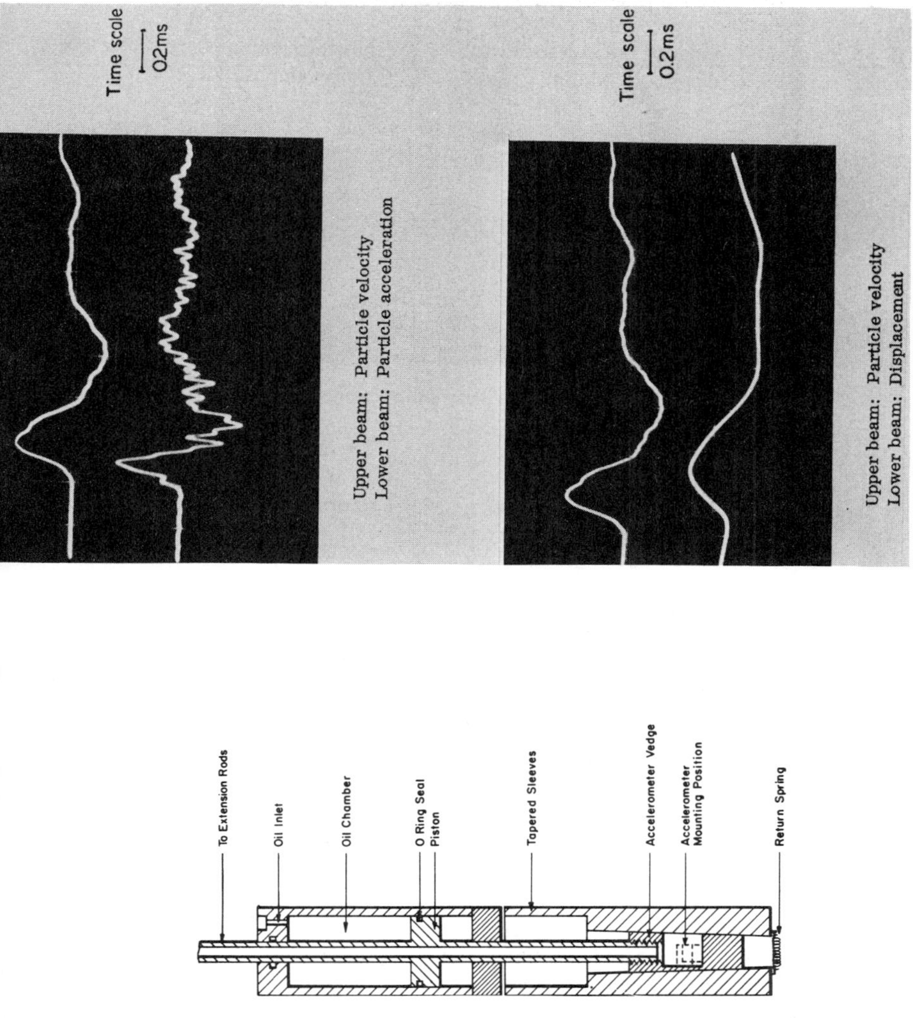

Fig. 14 —Accelerometer measurements during blasting.

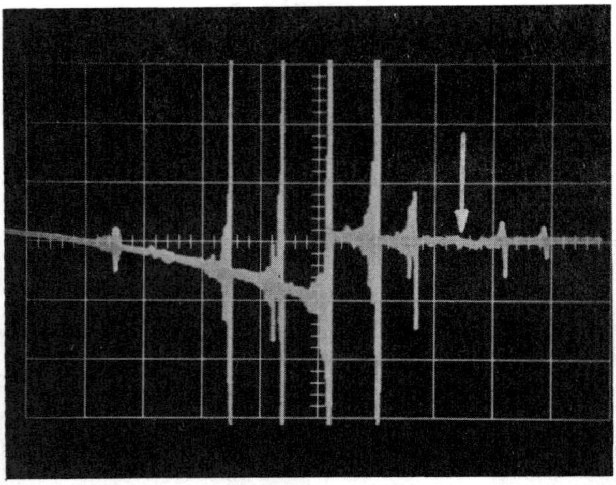

Fig. 15 —Acceleration record from production bench blast.

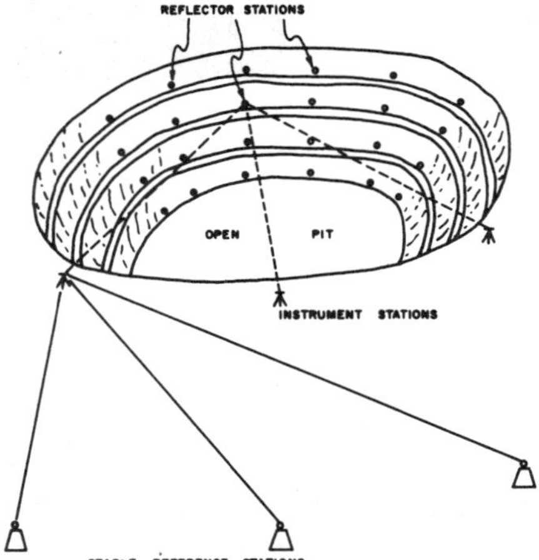

Fig. 16 —Diagrammatic layout of measurement network.

ever, it is possible theoretically to determine the complete stress tensor from measurements in a single hole provided strain or deformation measurements are made in six different directions; no more than three being allowed in any one plane and no more than two of the remaining ones in another plane. Two techniques (19, 20) using these principles are under development and promise reduced costs and increased efficiency for the future. One technique (19) may possibly be developed for use in deep vertical downholes which would be particularly useful for slope stability problems.

REFERENCES

1. Wylie, K.M., "Seismic Analysis", The Testing World, No. 22, Soil Test Inc., 1968-69, pp. 4-5.

2. Stears, J.H., "Evaluation of a Penetrometer for Estimating Roof Bolt Anchorage", USBM Report of Investigations 6646.

3. Coates, D.F. and Gyenge, M., "Plate-load Testing on Rock for Deformation and Strength Properties", Testing Techniques for Rock Mechanics, Am. Soc. Testing and Materials, Special Tech. Publ. No. 402, 1966, pp. 19-35.

4. Jenkins, J.D., "Laboratory and Underground Study of the Bearing Capacity of Mine Floors", Int. Conf. on Strata Control, Paris, 1960.

5. "Surveying Instruments", VEB Carl Zeiss, Jena Publication.

6. Untersee, V., "Investigation of Spatial Deformations in Power Dams by the Geodetic Method", Swiss Federal Topographic Survey, 1951.

7. Hedley, D.G.F., "Design Criteria for Multi-wire Borehole Extensometer Systems", 1st Can. Symp. Min. Surveying and Rock Deformation Measurements, Fredericton, October 1969.

8. Gray, W.M. and Toews, N.A., "Analysis of Accuracy in Determination of the Ground Stress Tensor by Means of Borehole Devices", Status of Practical Rock Mech., AIME, New York, 1968.

9. Obert, L., Merril, R.H. and Morgan, T.A., "Borehole Deformation Gauge for Determining the Stress in Mine Rock", USBM, RI 5978, 1962.

10. Coates, D.F. and Yu, Y.S., "Three Dimensional Stress Distributions Around a Cylindrical Hole and Anchor", 2nd Congress Int. Soc. Rock Mech., Belgrade, Yugoslavia, 1970.

11. Leeman, E.R., "A Trepanning Stress-relieving Technique for Rock Stress Measurement", Eng. and Min. Jour., Vol. 165, 1964.

12. Gray, W.M. and Barron, K., "Stress Determination from Strain Relief Measurements on the Ends of Boreholes; Planning, Data Evaluation and Error Assessment", Int. Symp. of the Determination of Stresses in Rock Masses, Lisbon, 1969.

13. Roberts, A., Hawkers, I.F.T. and Dhir, R.K., "A Laboratory Study of the Photoelastic Stress Meter", Int. Jour. Rock Mech. Min. Sci., Vol. 1, May 1964.

14. Barron, K., "Glass Insert Stress Meters", Trans. Soc. Min. Eng., AIME, December 1965.

15. Zauger, C.N., "Theory and Problems of Water Percolation", U.S. Bureau of Reclamation, Engineering Monograph No. 8, 1953.

16. Larocque, G.E., et al, "Field Blasting Studies", Proc. 4th Can. Symp. on Rock Mech. Symp., Queen's Printer, Ottawa, 1967.

17. St. John, C.M. and Thomas, T.L., "The N.P.L. Mekometer and its Application in Mine Surveying and Rock Mechanics", Trans. Inst. Min. Met., April, 1970.

18. Smith, K. and Oliver, E., "Computer Program for Determining Movement in Open Pit Slopes", Mines Branch Internal Report MR 69/67-ID, July, 1969.

19. Rocha, M., "Discussion on Theme 4: Residual Stress in Rock Masses", Proc. 1st Congress Int. Soc. Rock Mech., Vol. 3, Lisbon, 1966.

20. Leeman, E.R. and Hayes, D.J., "A Technique for the Determination of the Complete State of Stress in Rock Using a Single Borehole", Proc. 1st Congress Int. Soc. Rock Mech., Vol. 2, Lisbon, 1966.

Chapter 10

THE STABILIZATION OF SLOPES IN OPEN-PIT MINING

by

H. Q. Golder

Principal, H. Q. Golder and Associates Ltd., Toronto

INTRODUCTION

The stabilization of slopes is not a design problem. A slope is designed before it is made. For a slope to be stabilized it must already exist.

The basic problem has existed for a long time. It started with cuts in soil and rock made in the 18th century during the construction of the canals which were developed at that time, and continued into the 19th century with the development of the railways, and in the 20th century with motor roads and lately open pit mining.

It is not difficult to stabilize a slope, but certain questions must be asked and answered. These are:

i) How much stability is required?
ii) For how long is it required?
iii) How important is the cost?

These three questions indicate the basic problem of Civil Engineering, which is that the solution to any problem is a compromise between safety, cost and time. Stabilization of slopes is no exception.

If the slope exists why is stabilization a problem? There are many possible reasons, for instance, a neighbouring and similar slope may have failed, or it may be necessary to change the conditions of the slope, steepen it, increase its height, put weight at the top, or construct a building at the foot in which case added security is required, or guarantee its stability for a longer time.

MECHANISM OF FAILURE

In order to pick the correct method of stabilization the mechanism of failure must be understood and the immediate cause of failure known. Often to remove the cause is the best remedy.

The mechanism of failure has been studied over a period of many years in landslide problems. Pit slope failures are very similar to some types of landslides. (1) (2) (3) (4)

In one sense there are many causes of landslides. Terzaghi lists 19. This means that there cannot be a standard solution to a stabilization problem. In another sense, it can be said that there is only one cause, which is gravity. This is a constant, never failing force which cannot be turned off, and which ensures that anything loose moves downwards.

The principle of the mechanism is simple. There are disturbing forces, which are due to gravity, and resisting forces due to the strength of the soil and rock. Water is also involved; this has weight but no strength, and can exert a pressure due to gravity.

The factor of safety (F) is defined as the ratio of resisting forces to disturbing forces. If $F > 1$ the slope is stable.

The factor of safety can drop below unity due either to a reduction in the resisting forces or an increase in the disturbing forces.

The factor of safety is not constant but is subject to cyclic changes caused by variations in water pressures due to rain and melting snow. In the case of hard jointed rocks overlying softer rocks, creep in the softer rocks under the load of the harder rocks can cause the progressive rupture of blocks of the hard rock acting as dowels, so that there is a progressive reduction in the factor of safety due to this cause. Terzaghi explains the great landslide at Turtle Mountain near Frank, Alberta, in 1903 by this mechanism (see Fig. 1). In this example the disturbing forces varied cyclically, and the resisting forces dropped gradually. The slide occurred the first time the drop in F due to water pressure brought the value of F below unity.

To stabilize a slope, either the disturbing forces must be reduced or the resisting forces must be increased.

The value of 'g', the acceleration due to gravity cannot be changed, but the disturbing forces can be changed by altering the shape of the slope. Flattening the slope, or removing weight from the top, or putting weight on the toe in the form of a berm reduces the disturbing forces and increases the value of F. The resisting forces can be increased by increasing the strength of the soil or rock. The most effective way to do this is to drain water from the soil or rock since this increases the effective stress in the soil, or rock. Drainage also reduces weight

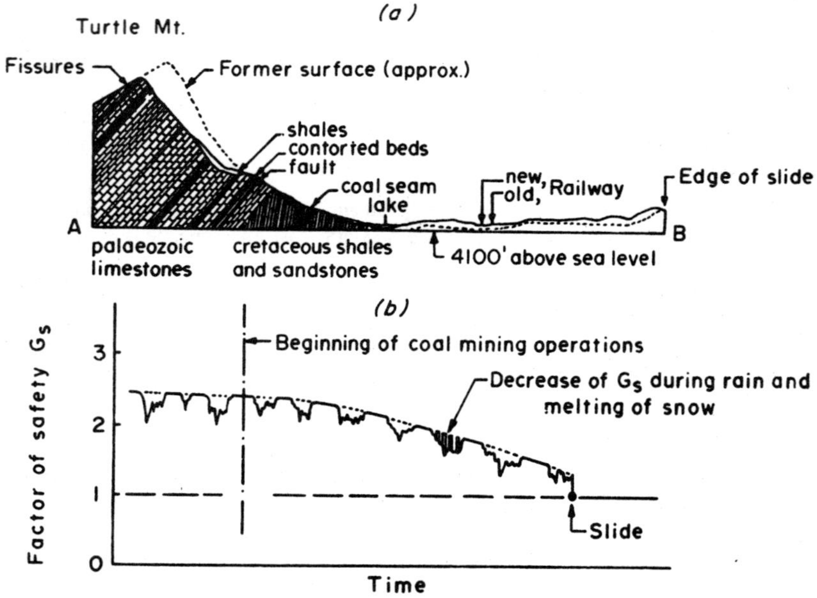

Fig. 1 —Variation of Factor of Safety with time, Turtle Mountain slide (1).

but this effect is small, and it removes cleft water pressures in rock which are a disturbing force. The strength of granular soils can be increased by compaction or by loading as by a berm.

It can be seen that some measures can increase F both by reducing disturbing forces and by increasing resisting forces. The resisting forces can also be increased by mechanical means such as rock bolting and grouting.

These are the only methods of stabilization which we have.

PIT SLOPES IN SOIL AND ROCK

Slopes of open pits differ in some respects from landslides. In general the slopes are comparatively steep and failure will be shear failure, so that some of the mechanisms of landslides such as creep and flow will not occur. The time scale is different since in an open pit stability is required for relatively few years and even in this time period the importance of a particular slope can change with time. But the same principles apply.

MATERIALS

Three materials enter into problems of slope stability: soil, rock and water.

In very arid areas water may play only a minor role, but in general, jointed rocks contain water in the joints, certainly after a rainstorm, and the pressures exerted by the water are significant.

The fact that no water is apparent on the face does not mean that there is no water in the joints. It takes very little water to exert a pressure. Drainage holes or adits made into an apparently dry face will often yield a surprising amount of water.

Although in some mines the whole face may consist of rock, in other cases there may be 300 ft. or more of soil overlying the rock. A failure in this overburden can be quite as disastrous as a failure in the rock. See example given by Brawner (5).

SLOPES IN SOILS

Quite a lot is known about the stability of slopes in soils and in the case of relatively homogeneous soils it is possible to design a safe and economical slope. The same methods are applied in the design of stabilization methods.

In the simple case shown in Fig. 2 a man-made slope of clay failed along a cylindrical surface of rupture. From the fact that at failure the factor of safety was unity it was possible to calculate the shear strength of the clay (S_u = 470 lb/ft.^{2}). Unconfined compression tests on samples of the clay taken from boreholes gave a shear strength of 500 lb/ft.^{2} This close agreement confirms both the method of analysis and the method of testing.

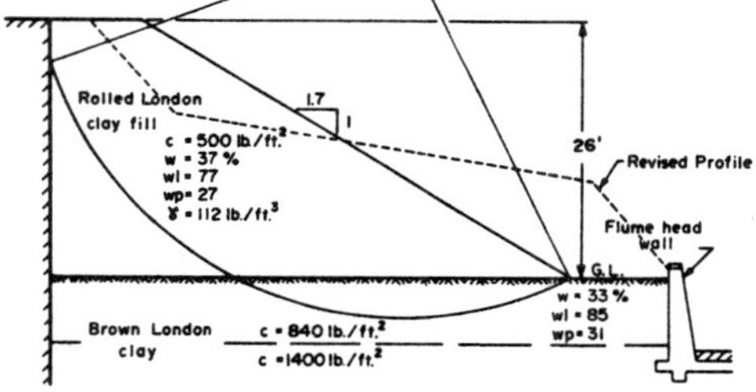

Fig. 2 —Stabilization of a slide in clay by changing the shape of the slope.

It was not possible to increase the shear strength of the clay and so the slope was stabilized by changing its shape as shown. Soil was removed from the top of the slope and placed at the bottom thus reducing the disturbing forces. From a simple analysis it was possible to show that the factor of safety had been increased by 25% to a value of 1.25.

The surface of failure in soils often approximates to a plane or a cylinder. On these surfaces the energy of deformation is small (or even zero), and such failures are easy to analyse.

When the soil is not homogeneous, the surface of failure varies from a plane or a cylinder and tends to go through the weakest material (see Fig. 3). In the case shown the slope was stabilized by removing some of the weak material and substituting a gravel toe (6). This had two effects, the gravel under its own weight formed a frictional key increasing the resisting forces, and the change in shape of the band reduced the disturbing forces.

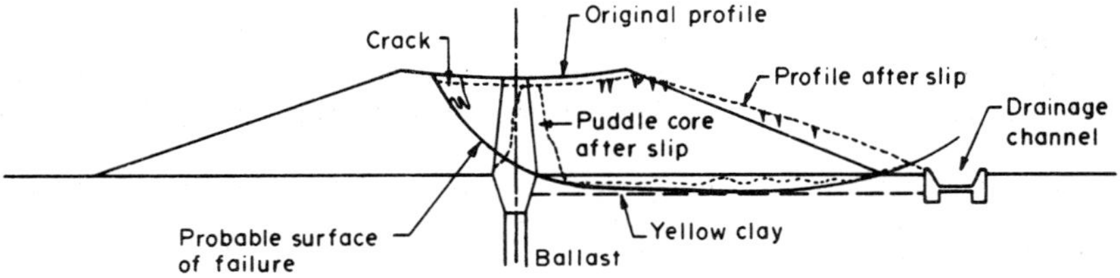

Fig. 3 —Slide through a non-homogeneous section.

Fig. 4 shows cliffs of a stiff-fissured clay which were continually failing due to erosion by the sea at the toe, softening of the surface due to weathering and infiltration of rain water into the fissures (7). The cliffs were stabilized by grading to a flatter slope, constructing a sea wall to prevent erosion and installing an elaborate drainage system as shown. The drains up the slope are 10 ft. deep and filled with stone. The chevron drains are 4 ft. deep. The effect of the drains is to prevent softening of the clay, remove water from the fissures, and lower the water table in the slope, thus increasing the effective stress and hence the shear strength.

SLOPES IN ROCK

Study of the problem of the stability of rock slopes is about 30 years behind that of soil slopes. The basic principles are the same — shear strength, effective stress, water pressure, — but the answers to the problems are by no means all known as yet.

There are two very large differences between slopes in rock and in soil.

The first is that compared to rocks, soils can often be considered almost homogeneous. In Fig. 3 the difference in strength between the soft clay and the stiff bank material is small compared to the difference in strength of a block of rock, and the strength along a joint, a bedding plane, or a fault. That is to say, that the behaviour of a rock mass forming a pit slope is controlled by the defects in the rock; angle of friction of rock on rock in a joint or along a bedding plane, or shear strength of gouge material in a fault, or infill material in an open joint.

The second difference is that we have not yet checked the theories of strength and failure mechanism against actual failures, and modified both the theories and the methods of measuring strength to give the right answer, which means that the calculations should show a factor of safety of unity for a failure. This we can do in soils.

Fig. 4 —Stabilization of cliffs by flattening the slope, constructing a sea wall and installing interceptor drains.

The 'back-analysis' referred to by Pentz (8) can give information on strength for a particular pit, which information can be immediately useful in designing stabilization measures. But this is only the first step. The goal must be the ability to measure the strength properties and predict stability and so decide on stabilization measures. This may take the next 30 years.

JOINTS IN ROCK

1. Fig. 5 represents a rock slope with an inclined joint or fault zone as shown. This is a plane of weakness and it is evident that if failure is likely to occur it will occur by sliding along this plane.

   The angle of friction for rock on rock along the joint, or for the joint filling material can be measured in laboratory tests. If the weight of the rock and the water pressures in the joint system are also measured then the stability of this system can be analysed quite easily. If the stability proves to be inadequate how can this slope be stabilized?

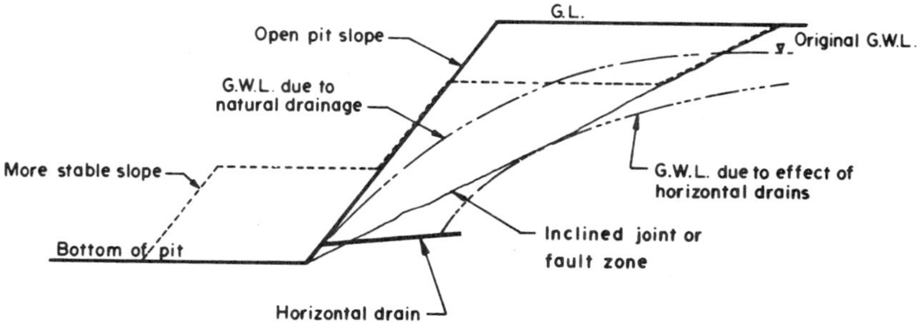

Fig. 5 —Typical methods available to stabilize an open pit slope which contains an inclined joint or fault zone.

As stated above there are basically only three methods:

a) the disturbing forces can be reduced. This can be done by removing rock from above the joint plane at the top of the slope and building a berm at the toe. The berm also increases the resisting forces.

It is obvious that to remove all the rock above the joint plane solves the problem. It is not so obvious, but none the less true, that to remove rock from the top of the slope only does not increase the safety of the slope unless this action also reduces water pressures in the joint.

b) the shear resistance of the rock along the joint can be increased by draining water from the joint and so increasing the effective stress on the joint. This is only so if there is water in the joint and if it can be removed.

c) the resistance can be increased by artificial means such as rock bolts through the joint. The shear resistance of the bolts alone is not great but if the bolts are tensioned a considerable increase in resistance results due to increased friction forces. In this example the plane of rupture is known so that one knows where to drain and where to bolt.

2. A common joint pattern in sedimentary, horizontally bedded rocks is shown in Fig. 6. In a steep slope, there will often be an open vertical joint near the top of the slope. Below this, if the joints and bedding planes are closely spaced, and if the water level is high a stepped failure surface consisting of a series of vertical joints with zero strength and full of water and short horizontal steps with frictional strength only, can develop. This surface approximates to a portion of a cylinder and the amount of crushing of intact rock necessary to permit movement is minimal.

The surface of failure is not known beforehand and must be found by calculations involving a process of trial and error.

The most effective stabilization method in this case is drainage. If the water level can be lowered as shown the shearing resistance can be increased.

3. In many rocks, particularly igneous rocks, the orientation of the joints is varied and at first sight may appear to be completely haphazard. But measurement of joint orientation often reveals a three dimensional pattern to the jointing with one or more preferred directions of strike and magnitude of dip.

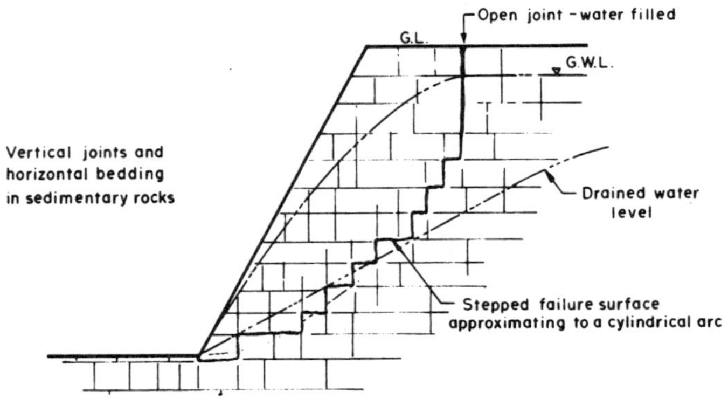

Fig. 6 —Common joint pattern in sedimentary, horizontally bedded rocks.

This pattern must be observed and recorded in order to design a safe slope. In many cases boreholes or adits used for joint surveys can be used to aid drainage and so increase stability. They can also be used for the installation of measuring instruments since it is not enough to initiate stabilizing actions, one must also show that they have been effective in increasing stability.

PROCEDURE USED IN STABILIZING A SLOPE

The first question to be decided is the degree of urgency.

If the slope has started to move then the degree of urgency is high. Normally one would only know this if the slope was instrumented but cracks in the ground and breakage of water mains and sewers can be considered inadvertent instrumentation.

The first action is to prevent loss of human life. If inhabited dwellings are in danger then immediate evacuation of the people is the proper course.

The next action is to cut off the entry of water from streams, ditches, broken water mains and melting snow into the slide area. Surface water and rain water must be led away from the top of the slide by ditches or other means. Any surface cracks at the top of the slide area whether in soil or rock should be sealed with clay or other suitable impervious material (polythene sheets over the area will help).

Drainage measures should next be started to lower the water level in the unstable mass of soil or rock. These can be horizontal pipe drains drilled into the face of the slope, or adits or trenches. Anything which causes water to run out of the slope is a help. If water is already running out of the face of the slope, a pipe drain at this point will probably increase the rate of flow. But areas where the face is dry should not be neglected. A pipe drain in such an area may pierce a steeply inclined impervious fault zone and yield abundant water. A dry face may mark a higher pressure than a wet face, and the purpose of drainage is to reduce water pressure. If possible unload the top of the slope.

The urgency can only be diminished by instrumentation which shows that the measures taken are having a beneficial effect and the stability of the slope is increasing. So that as soon as possible after the initial urgent measures have been taken, or at the same time if the personnel and equipment are available, piezometers, movement hubs and possibly inclinometers should be installed.

It is often found, particularly in soils or weathered rocks that a plot of water level in piezometers against rate of movement of the soil or rock mass will yield a straight line. If this is projected back to zero rate of movement a water level is indicated, which when reached by drainage, will ensure at least temporary stability and allow time for further study.

Once the urgency is over it is necessary to determine the cause of instability, by visual observation based on experience, by water level measurements, by slope indicators, by tests on the materials, and by a survey of the joint pattern. Laboratory tests are required on soils and samples of rock adjacent to joints, and above and below bedding planes, and on joint infill and gouge in fault zones.

A decision must then be taken on how the slope is to be stabilized, which method is to be used and what instrumentation is required to demonstrate that the required increase in stability has actually occurred.

METHODS OF STABILIZATION

Basically there are only three methods available, which are:

1. Mechanical strengthening
2. Change shape of slope
3. Drainage

In most cases the choice of method is based on cost.

1. **Mechanical Strengthening**

Mechanical strengthening can be accomplished by rock bolting, stressed steel anchors (either bolts or cables) and grouting.

Grouting

A fissure or an open joint can be grouted, but there is no increase in strength until the grout has set. Even after the grout has set there will be little increase in strength if the joint faces are lined with soft material which is not displaced by the grout. This is often the case.

Before the grout has set it is a heavy fluid and exerts a pressure which decreases stability. This means that a fissure close to an open cut must not be grouted unless the face is previously bolted into the mass of the rock. For this reason grouting is of limited value in rock slope stabilization.

Rock-bolting

This is a very generally used method in tunnels in rock to hold up the roof in the zone which has been loosened by blasting and excavation. Steel bolts 10-20 ft. long are inserted into percussion drilled radial holes in the roof and held by an expanding foot at the end of the hole, or alternatively by grouting the steel into the hole. A small amount of pre-stress is put into the bolt by the nut at the outer end.

This method can be used to stabilize slopes, but as shown below, in general a greater amount of pre-stress than is normally used with bolts is required for efficiency.

Pre-stressed cables

In principle these are the same as rock bolts but it is possible to apply considerable pre-stress in the cables. This increases the normal pressure across a joint or bedding plane and so the frictional resistance.

The technique has been used in Civil Engineering for many different purposes.

High tensile steel cables are grouted into holes drilled into rock and then stressed using jacks. The effect is to increase the normal pressure across a joint and so to increase the frictional resistance. This increases F.

If the joint is filled with an infill material with low ϕ value then the efficiency of the method is very much reduced. But for rock on rock where $\phi > 30°$ the method is very efficient.

The principle is simple. In Fig. 7 the block of weight W is stable if $\alpha < \phi$ since

$$F = \frac{\text{resisting force}}{\text{disturbing force}} = \frac{\mu W.\cos\alpha}{W.\sin\alpha} = \frac{\tan\phi}{\tan\alpha} \qquad \text{........ (1)}$$

where μ = coefficient of friction

This is independent of W.

So that F cannot be changed by changing W.

But if the block of weight W is tied to the sloping surface by a force P by stressed cables, W has not been increased but the resisting forces have and now

$$F = \frac{\mu W.\cos\alpha + P}{W.\sin\alpha} \text{ which is } > \frac{\tan\phi}{\tan\alpha} \qquad \text{........ (2)}$$

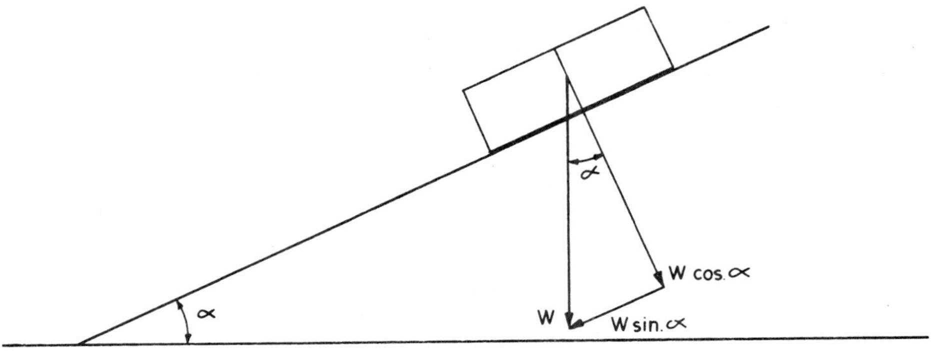

Fig. 7 —Resolution of forces of a block on a slope.

So that F has been increased.

This device is useful but costly.

2. **Change shape of profile**

It is obvious that a flat slope is more stable than a steeper slope. What is not so obvious is that a bermed slope as shown in Fig. 5 is more stable than a plane slope.

One point which must be watched here is that the stability of the steeper but lower slopes formed by the shaped slope is adequate.

Under change of shape can also be included removal of rock above a potential surface of failure. This also can increase stability if the rock is removed from the right place.

3. **Drainage**

Generally drainage is a most effective method of increasing stability. However if there is no water in the fissures of the rock then drainage will be ineffective.

That there is no water in the pit does not necessarily mean that there is no water in the rock. As stated above, a fault zone filled with impervious gouge can create a zone of high water pressure but no seepage. Further, where the seepage is small, in a hot climate evaporation can remove all water at the rock surface, but nonetheless a water pressure may exist in the rock mass.

In most areas there is water in the ground at some distance below the surface. When a pit is made the horizontal stresses in the ground are relieved and vertical and inclined joints open and water seeps into these joints. This has two related effects:

i) a horizontal water pressure is set up in an open vertical joint. This is a disturbing force.

ii) a water pressure is set up around any potential surface of failure. The pressure is perpendicular to the surface of failure and reduces the effective pressure on this surface, and so the shear strength and the resisting forces and the factor of safety.

The quantity of water required to create a pressure is very small in a small fissure.

Drainage removes or reduces these water pressures.

METHODS OF DRAINAGE

If a pit is excavated, drainage will occur naturally and the previous water level will drop towards the face of the slope as shown in Figs. 5 and 6, and water will issue from the face of the slope at a certain level. To improve stability it is required to reduce the water level to something like the dashed line in the figures.

There are several different methods of doing this which are discussed below.

Wells

A ring of deep wells can be constructed around the pit from ground level and going down to or below the bottom of the pit (Fig. 8). Water is pumped continuously from these wells. The water table is lowered as shown.

This method is not cheap, but if the wells are put in before excavation is commenced it will be possible to excavate to steeper slopes than would be possible without drainage. This saving can sometimes pay for the wells.

One great advantage of deep wells is that they do not freeze up in the winter.

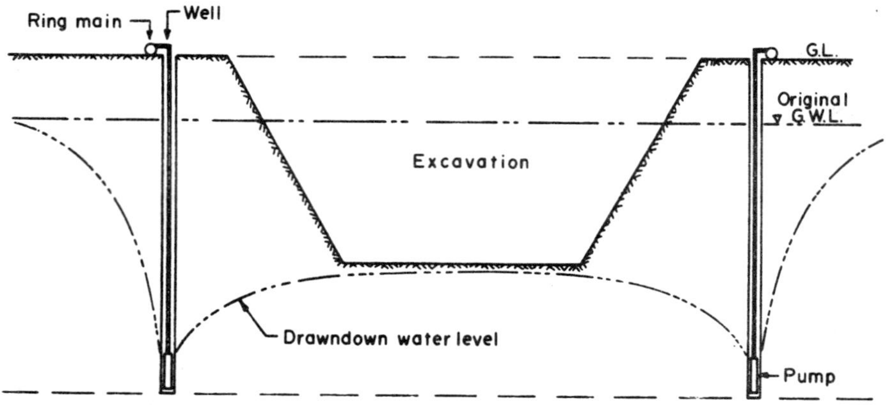

Fig. 8 —Drawdown of a water table by wells.

Horizontal Drains

In an existing slope it is possible to install horizontal pipe drains into the slope from the foot. These drains will bleed off water from behind the slope, lower the water level and increase stability.

The technique has been used successfully many times — particularly in highway cuts, over the last ten or more years.

The drains consist of perforated steel pipes 2 in. - 3 in. diam. drilled in horizontally, at a few degrees above the horizontal to distances of 200-300 ft. from the face.

If it is important that the drains should continue to flow during the winter in low temperatures then the discharge must be into a covered and possibly insulated duct.

Adits

Adits driven into the rock near the base of the slope can be a most efficient form of drainage if they are correctly positioned.

Adits are expensive, but if they can be made to serve two purposes, drainage and exploration, the expense can often be justified.

An adit can be used for drainage and also for a detailed joint survey, for which purposes it is much more effective than a borehole, or even a number of boreholes. An adit can also give information on the variation of permeability which is of great importance. This information is obtained by measuring the flow of water separately from each section of the adit.

In an adit there is no freezing danger provided simple insulation precautions are taken.

Recent original theoretical work on this problem shows that a normal adit size of an equivalent diameter of 7 ft. gives good drawdown for a slope 300 ft. high (9). These figures depend on the ratio of k_h/k_v . As this ratio increases the efficiency of the adit can be increased by vertical and inclined drill holes from the adit, which can be used to puncture impermeable zones. The position of the adit is of great importance from the point of view of efficiency of drainage.

When adits are used there is a strong case for spending a little more money to measure everything, i.e. quantity of water flowing at different points along the adit, and temperature, quality and chemistry of the water. All this is in addition, of course, to detailed joint surveys.

MONITORING AND FIELD MEASUREMENTS

When a stabilization measure is started, the success of the operation is not known. The operation (which costs money) is based on an engineer's appraisal of its effect on the stability of the slope. The effectiveness can only be determined by measurement of simple things like movement and water pressures — including rate of change of these factors.

Money spent on measurement is rarely wasted — provided the program of measurement is decided by the engineers in charge of the job. Further this means that the same personnel must be used throughout the job. An organization, in which one team drills a hole, a second team installs a slope indicator, a third team takes readings, and a fourth team interprets the results and deals with the engineering problem, has no part in the solution of stabilizing problems in pit slopes.

What one does in measurement, and the extent in money to which one goes depends on how close to a factor of safety of unity one is working.

One measures

a) water levels — there may be more than one

b) quantity of water flowing from drains and adits or pumped from wells

c) movements of surface — both horizontal and vertical at ground level, along the slope and at the pit floor

d) movements within the mass — i.e. slope indicators in boreholes

e) change in size of cracks — at the surface and in adits

f) change in diameter of adits — vertical and horizontal

g) permeability in boreholes — by falling head or other tests in soil — and by packer tests in rock.

THE SOLUTION TO A PARTICULAR PROBLEM

Each job is unique. Problems recur on successive jobs — but they are never quite the same.

This is the charm of our discipline.

On every job one MUST start from first principles — one investigates — one measures — one experiments a bit — and makes a decision. That is engineering — there is NO MAGIC.

Though presidents of companies and shareholders may demand — they cut no ice against Nature and Gravity — they are the masters in these problems — first to understand — then to obey, and then to control within the understood and inexorable limits.

We can operate intelligently within the bounds of natural law, outside of these bounds, there is NO MAGIC.

REFERENCES

1. Terzaghi, K., "Mechanism of Landslides" Berkey Volume, Geological Society of America, Engineering Geology (Berkey) Volume 1, pp. 83-123, 1950.
2. Sharpe, C.F.S. "Landslides & Related Phenomena" Columbia University Press. N.Y. 1958.
3. Highway Research Board, "Landslides & Engineering Practice" NAS-NRC Publication 544, Washington, D C., 1958.
4. Skempton, A.W. & Golder, H.Q. "Practical Examples of the $\phi = 0$ Analysis of the Stability of Clays" Proc. 2nd Int. Conf. S.M.F.E. — Rotterdam. Vol. II. p. 63., 1948.
5. Brawner, C.O., "Case Studies of Stability on Mining Projects," Symposium on Stability for Open Pit Mining, Vancouver, B. C., 1970.
6. Cooling, L.F. & Golder, H.Q., "The Analysis of the Failure of an Earth Dam during Construction" Jour. I.C.E. Nov., 1942.
7. Duvivier, J., "Cliff-Stabilization Works on London Clay" Jour. I.C.E. Oct., 1940.
8. Pentz, D.L., "Methods of Analysis of Stability of Rock Slopes" Symposium on Stability for Open Pit Mining, Vancouver, B. C., 1970.
9. Sharp, J. C., "Drainage Characteristics of Subsurface Galleries" Proc. 2nd Int. Congress on Rock Mechanics, Belgrade, 1970.

Chapter 11

DESIGN AND CONSTRUCTION OF TAILINGS DAMS

by
Leo Casagrande

Professor of the Practice of Foundation Engineering, Harvard University

and
Bruce N. McIver

Partner, Casagrande Consultants, Arlington, Mass.

ABSTRACT

Similarities are noted in the practices and problems of constructing dams to impound mine tailings, fly ash, and chemical wastes. The comparison of a typical tailings dam with a conventional water-retaining earth dam shows that the former is inherently less stable and can be susceptible to cracking, liquefaction, erosion, and frost action, as illustrated by examples. Several methods are presented for building safe tailings dams using either the coarse fraction of the tailings or local soils, the important factor in each method being proper compaction of the material. The safety of existing tailings dams can be improved by flattening the overall slope, preferably by adding berms, or by decreasing the height. Where neither remedy is possible, the installation of electro-osmotic tie-backs is suggested. Dumps of mine spoil can be stabilized economically by compacting the material in layers along the toe to form a dam.

INTRODUCTION

The problems encountered by the mining industry in the design and construction of tailings dams are very similar or even identical to those arising from the impounding of fly ash by electric power companies or of wastes from chemical industries. With only a few exceptions, all of these waste materials have a high percentage of grain sizes passing No. 200 mesh. Mine tailings and fly ash with 80% passing No. 200 mesh and chemical wastes with up to 100% smaller than No. 200 mesh are not uncommon.

Apart from recent experimental efforts at storing fly ash in the dry state, practically all waste materials are being sluiced into storage basins. After settling out in the pond, these waste deposits have as a rule very high water contents and permeabilities which may range from 1×10^{-3} cm/sec. to 1×10^{-8} cm/sec., depending on the distance from the discharge points. The coarser types of mine tailings may consolidate at the same rate fresh tailings are being added, but chemical wastes and fine fly ash frequently develop a thin highly impervious surface crust which may appreciably delay the rate of consolidation and, as a result, cause excessive pore pressure. However, even after consolidation is completed, waste deposits remain sensitive to liquefaction (1 to 5) or straining by oversteepening of a slope.

COMPARISON OF TAILINGS DAMS WITH CONVENTIONAL EARTH DAMS

Considering the similarities of the problems involved in storing mine or industrial wastes, it is perhaps not surprising to find that the methods of constructing retaining dikes for all wastes have developed along similar lines. First, an earth dike of normally 10 to 20 ft. in height is built. When the storage pond approaches this height, additional dikes are added on top as shown in Fig. 1. Except for the original earth dike, all subsequent dikes rest in part on the dike below and in part on the surface of the tailings pond. Where mine spoil becomes available from strip mining, it is sometimes used for the construction of these dikes. If mine spoil or other suitable fill material is not available at a reasonable cost, the dikes are constructed by scraping from the pond surface the thin desiccated crust of mine tailings. Chemical wastes, too, are often used in the same manner to gradually build up the dam. A typical mine tailings dam which grew gradually to over 300 ft. in height by simple hand operation is shown in Fig. 2. Another gigantic tailings deposit in the gold mining area of South Africa is shown in Fig. 3.

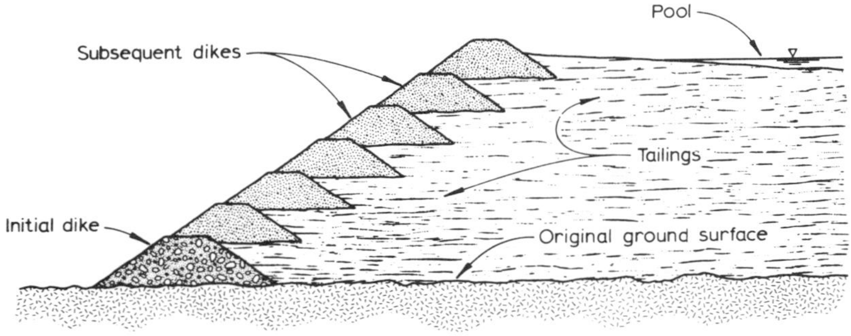

Fig. 1 —Section through typical tailings dam

Fig. 2 —Growth of tailings dam in arid region

Fig. 3 —Tailings dams in South Africa

Fig. 4 shows a comparison between the cross-section of a typical tailings dam and the cross-section of a conventional earth dam for impounding water. Such a comparison, together with consideration of the design and construction practices used for each type of dam, leads to the following observations:

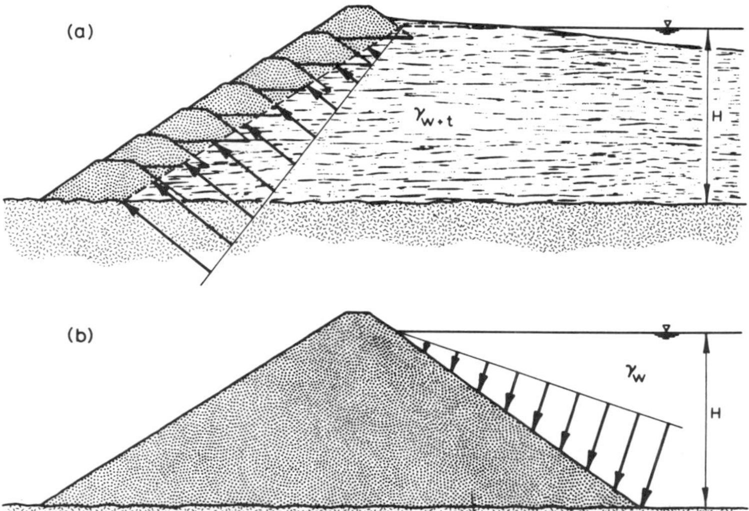

Fig. 4 —Pressure acting on (a) tailings dam and (b) conventional earth dam

1. The earth dam is built of much stronger materials than the ponded wastes and is constructed in thin, well-compacted lifts. In contrast, the ponded wastes are very loose and have a high water content and appreciably smaller shear strengths than the materials used for a conventional dam.

2. If a conventional earth dam which retains impounded water is designed with a factor of safety of 1.5, the factor of safety of a "shell-type" dam impounding saturated tailings with the same slope would obviously be very much smaller, as can be seen in Fig. 4.

3. A properly compacted dam cannot liquefy even under severe seismic activity. In contrast, loose and saturated granular or chemical wastes deposited behind a relatively thin shell of supporting material could cause failure of the tailings dam. While undisturbed tailings may adequately contribute to the stability of the dam, the strength of such a "shell" cannot possibly withstand liquefied tailings.

A further reason why conventional tailings dams are more susceptible to trouble is that any cracks which may develop perpendicular to the axis may be more critical than in conventional earth dams. The development of cracks in earth dams perpendicular to the axis has been studied for some time (2, 6, 7) and designs are frequently modified in order to prevent such cracking or reduce its development to a minimum (8, 9). In the case of a thin shell-type tailings dam, the danger of cross-cracking is not only much greater but prevention is often not feasible. Because of the compressible nature of the loose waste material, differential settlements will be pronounced between section a-a and section b-b in Fig. 5 of a tailings dam across a valley with steep abutments. This results in tension cracks developing at locations of largest strain.

Another cause of perpendicular cracks is inherent in the areal configuration of some waste ponds. Since the vertical and horizontal stresses at sharp corners of tailings dams such

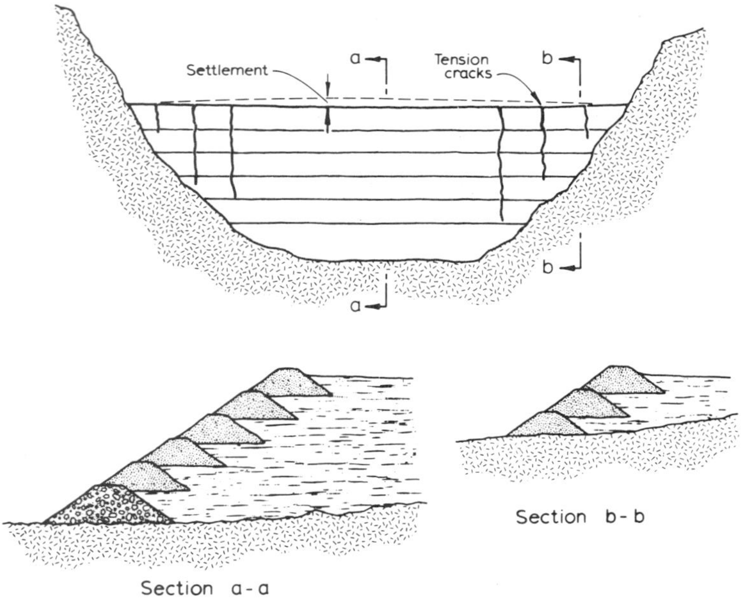

Fig. 5 —Development of tension cracks in dam by differential settlement

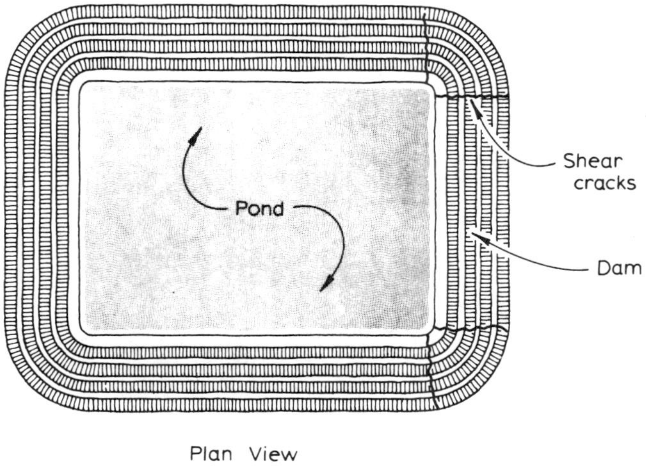

Fig. 6 —Development of shear cracks in dam by differential horizontal yielding

as shown in Fig. 6 are smaller than along the straight sides, an increase in height of the waste pond will cause less horizontal movement at the corners of the dike than along the straight sides. Such differential horizontal yielding can produce cracks which in turn may contribute to failure of a tailings dam. It is believed that such cracks may have contributed to the liquefaction failure shown in Fig. 7.

The sensitivity of wastes to liquefaction is indicated by the two consolidation tests shown in Fig. 8. A sample taken from the surface of a 100 ft. high waste pond which had been subjected to surface drying was tested both in the undisturbed state and after remolding. The starting water content for both test specimens was approximately 125% (void ratio e=2.9). Remolding and transferring the specimen into the consolidation ring and the consolidation under its own weight caused reduction in water content to about 110% (e=2.6) as revealed by the lower curve in Fig. 8. This illustrates that while a 100-ft. high undisturbed pile of submerged material may consolidate very little, after having flowed out it will consolidate into a much denser state. As soon as this material has time to reconsolidate after failure, it gains sufficient strength to be excavated with steep slopes without failing as shown in Fig. 9.

DESIGN CONSIDERATIONS

Where there is ample storage space, mining companies have for some time been using flat average slopes for their tailings dams by setting back subsequent dikes or groups of dikes as shown in Fig. 10. This flattening will obviously produce a much safer dam than the scheme discussed earlier. Although an average slope of 1 on 4, as established by the dashed line in Fig. 10, may permit an appreciable increase in the height of such a tailings dam, this in itself is not necessarily a guarantee that failure will not occur. The vibrations of a mine railroad transporting and dumping fill for such a dike, which had been set back a safe distance from the previous dike, may result in liquefaction of the underlying tailings as appears to have been the cause of the trouble shown in Fig. 11.

In a region where heavy rain or frost action or both will occur, tailings dams should not be built of fine mine tailings or fly ash. Such a practice could require corrective measures so costly as to completely outweigh the savings which might have accrued by using mine tailings

Fig. 7 —Typical liquification failure of waste pond

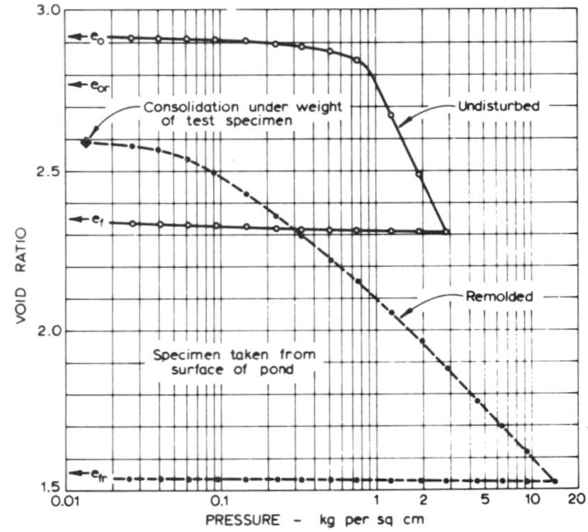

Fig. 8 —Consolidation curves for undisturbed and remolded waste material

Fig. 9 —Character of waste material upon consolidation after liquification failure

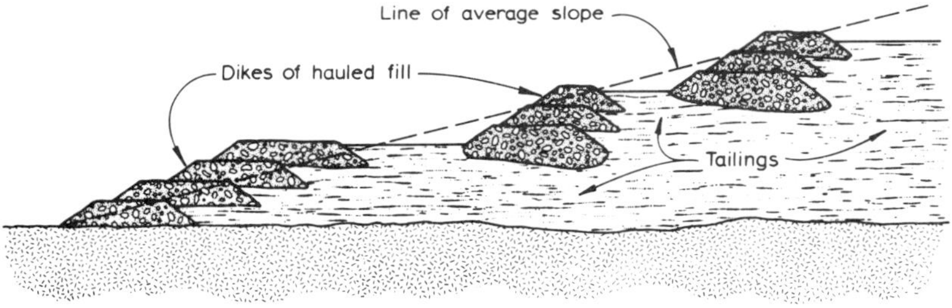

Fig. 10 —Improvement in stability by flattening overall slope

Fig. 11 —Dike failure by local liquefaction of tailings

rather than a more stable non-frost susceptible material. If seepage daylights along a slope of fine tailings, erosion of the tailings moves backward, as shown in the sketch in Fig. 12, until piping failure occurs. If frequent inspection catches such a situation at an early stage, as was the case with the tailings dam shown in Fig. 13, corrective measures are still possible. In this instance, piping had progressed within one day practically through the entire width of the dam and serious trouble was prevented only at the last minute. Once the "pipe" has short-circuited into the pond, as is the case in Fig. 14, control measures may no longer be effective.

If freezing blocks the drainage through the dike, and the tailings are susceptible to the growth of ice layers, as shown schematically in Fig. 15, thawing of such ice sheets in spring may

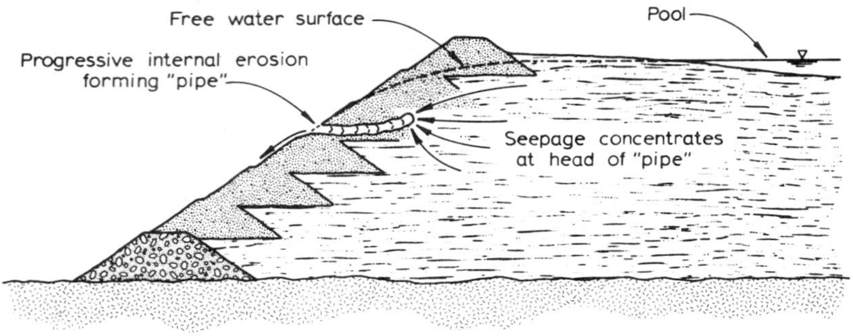

Fig. 12 —Development of pipe in tailings dam

Fig. 13 —Pipe developed in tailings dam

Fig. 14 —Advanced stage of piping

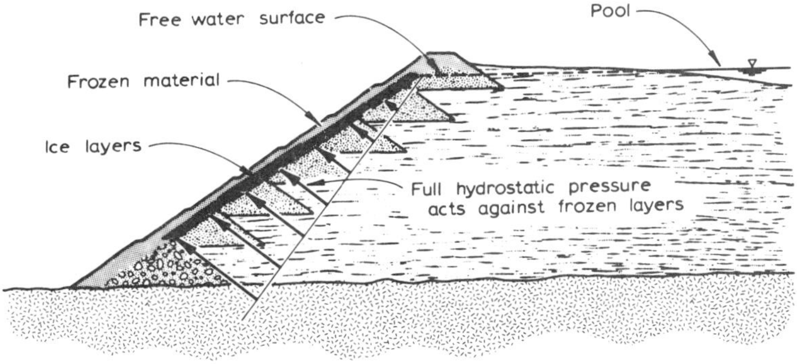

Fig. 15 —Growth of ice layers in tailings dam

lead to extensive sloughing, such as developed in the case shown in Fig. 16. In this instance, a number of "pipes" developed partway across the dike during the first few days of spring thaw.

Erosion starting on the outer slope can be controlled and corrective measures are often accepted as part of routine maintenance. Such maintenance, however, is not possible when erosion originates within the dike as was the case in the failure shown in Figs. 17 and 18. In this instance, the return pipe from the pond to the plant had been placed within the dike itself. Inevitable minor differential settlement resulted in a rupture of the pipe and rapid erosion of the dike. This occurred unnoticed overnight in spite of the excellent character of the dike material which consisted of a mixture of well-compacted clay shale and sandstone. The purpose

Fig. 16 —Sloughing of tailings dam upon thawing

Fig. 17 —Failure of tailings dike by internal erosion

Fig. 18 —Close up of dike after failure

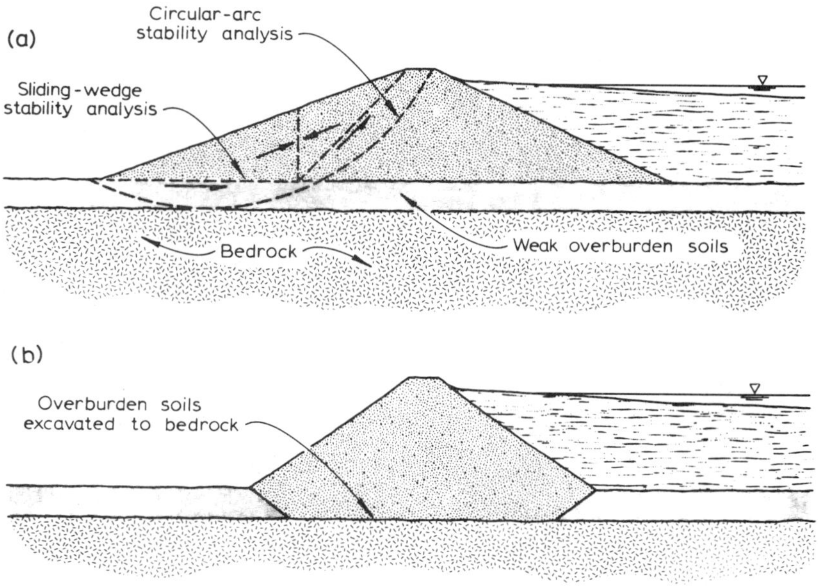

Fig. 19 —Comparison of cross section of dam on (a) weak and (b) strong foundation

of burying the pipe in the dike was to protect it against freezing and to effect savings by not having to raise the pipe with future increase in height of the dike.

From these foregoing sketchy examples it becomes apparent that the design of tailings dams and stability analyses of existing tailings dams can be more complicated than stability analyses of conventional earth dams. Once the strengths of a dam and of its foundation are known, the analysis is relatively simple. Whether to build a conventional earth dam with flatter slopes on the ground surface (Fig. 19a), or whether to excavate the overburden down to bedrock, to permit much steeper slopes (Fig. 19b), is simply a question of economics.

In a recent paper (10), an attempt was made to establish permissable heights and slopes for mine tailings in South African gold mines deposited by sluicing. Although in this instance the quartzite rock flour consists of 50 to 95% finer than 0.06 mm. and 5 to 15% finer than 0.002 mm., it has, due to its angularity, a relatively high angle of internal friction. The results of these investigations, assuming in one case an impervious base and in the other a completely pervious base, and for a factor of safety of 1.5, are shown in Fig. 20. Thus, for a material which is saturated and loose, and assuming a 10 ft. per year increase in height of the tailings, a 120 ft. high slope of 30° would be safe. If we were to establish similar rules of design for other surroundings and climatic conditions, or if earthquakes had to be taken into consideration, they would not nearly be as favorable as those in Fig. 20.

The answer to the question of how to build safe tailings dams and how to improve on the safety of existing tailings dams cannot be generalized. However, depending on the local conditions, the following methods of design and construction should be considered.

Use of Tailings for Construction of Dam

Except in a desert climate, where one can be certain that the tailings do not remain

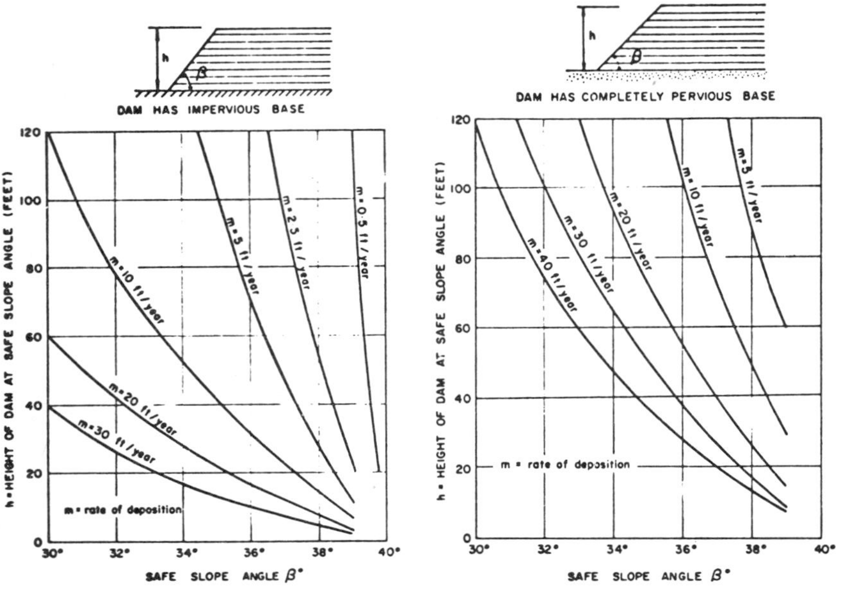

Fig. 20 —Theoretical design graphs for tailings dams in South Africa (10).

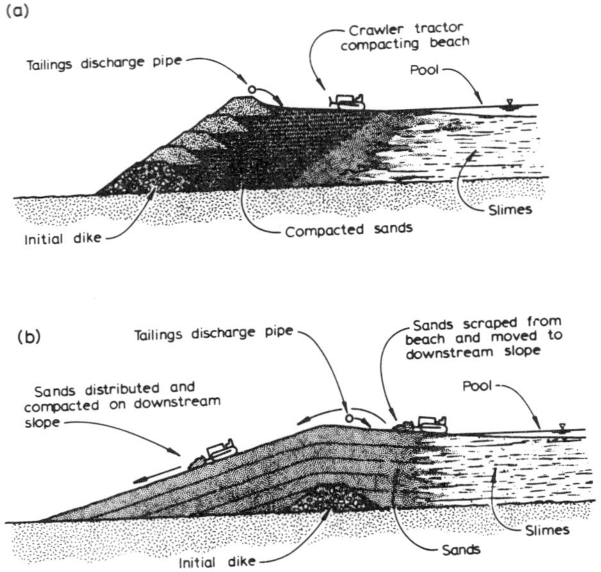

Fig. 21 —Methods of constructing dams of coarse fraction (sands) of tailings saturated, or under permafrost conditions where the tailings would remain permanently frozen, dams should not be built of fine tailings. However, tailings containing a reasonable percentage of sand sizes can be utilized if the discharge along a dike is arranged in a manner such that a beach is developed as shown schematically in Fig. 21a, and with the fines being deposited in a pool at a safe distance from the dam (11, 15). The sand along the beach should be spread and compacted with a number of coverages by crawler-type or rubber-tired tractors.

Instead of increasing the height of a tailings dam by adding new dikes on top of the bench, the sand can be moved over the crest to the downstream slope and there compacted with suitable equipment as illustrated in Fig. 21b. As a rule, such slopes are flatter than if developed by means of individual dikes. Fig. 22 illustrates a modification of this method with the tailings dam developing on the downstream side only (5, 13, 14, 15, 16).

Depending on the local conditions and on the relative quantity of sand sizes in the waste material, the separation of sand from the fines may be achieved either with cyclones or by sluicing using either side windows or bottom windows. Open pipe discharge with or without bottom windows has been used successfully.

Possibly one of the first tailings dams built by compaction of hydraulically deposited sand is the refuse disposal dam for a river coal plant in Pennsylvania which was started in the early 1950's and has by now reached a height of 135 ft. As indicated in Fig. 23, the original

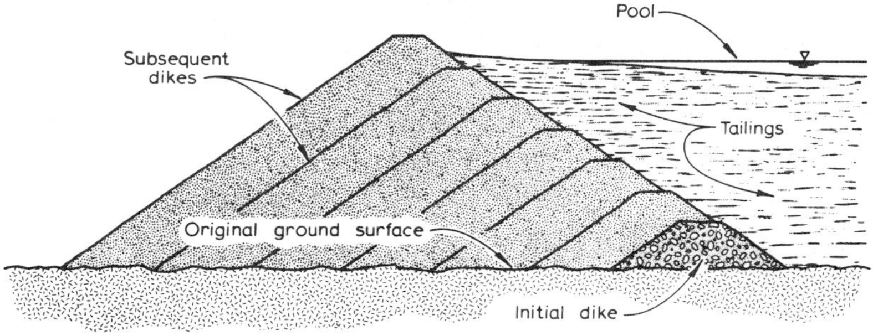

Fig. 22 —Modification of "downstream" method of tailings dam construction

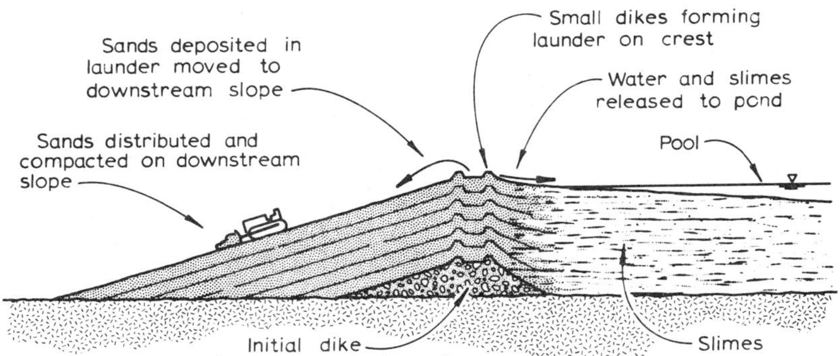

Fig. 23 —Construction of coal refuse dam in Pennsylvania

approximately 40 ft. high portion of the dam was built of local soil materials in compacted layers. The crest of the dam slopes gently from one abutment to the other. Along both edges of the crest, small dikes retain the coarser fractions of the refuse which is sluiced to the high end of the crest using an open pipe. At the opposite low end of the crest, the upstream dike has an opening to permit the fines in the effluent to discharge into the pond. The sand deposited along the crest, ranging in grain size from approximately 0.1 to 0.5 mm., is moved to the downstream slope by means of scrapers and dozers. Figs. 24, 25 and 26 show the sluicing operation, the pushing of sand from the crest to the downstream slope, and its compaction, respectively. The degree of compaction is checked with a simple hand penetrometer. Using very conservative assumptions, the factor of safety of this dam was computed to be at least 1.5.

Fig. 24 —Sluicing with open-end pipe for construction of coal refuse dam (note confining dikes along both sides of crest)

Fig. 25 —Moving sluiced material to downstream slope

Fig. 26 —Compacting material on downstream slope

A more recent example of a very large tailings dam exists in Alberta where the disposal of 125,000 tons of sand tailings per day is required in connection with the Great Canadian Oil Sands. This tailings dam, which was designed under the guidance of R. M. Hardy (12), has a length of several miles and may ultimately reach a height of 200 ft. Originally, the tailings were discharged from the open end of a 24 in. diam. pipe. However, because of difficulties in stabilizing the fresh tailings fast enough, this scheme was later modified by adding a "spoon", i.e., several feet of an open half-section of a pipe with several baffles and bottom windows as shown in Fig. 27.

Use of Mine Spoil or Local Soils for Construction of Dam

If adequate amounts of mine spoil or of other soils are locally available, and if construction of a dam can be started well ahead of the sluicing operation, this may offer the following advantages:

1. The dimensions and stability of the dam, including foundation and abutments, are established without the need for future efforts for improvement work.
2. Because the reservoir is not filled immediately, there is time for the dam and foundation to consolidate and to develop adequate strength. Therefore, the design can be based on a smaller factor of safety than the customary 1.5. If failure conditions should develop, they can be controlled by adding berms along one or both sides of the dam. A typical case where this method was used is the over 100 ft. high fly ash storage dam shown in Fig. 28.
3. Future increases of the height of the dam by stage construction can be considered during the design of the first stage. In recent years, a number of such dams for fly ash storage have been built and additional projects are in the design stage.

METHODS OF INCREASING THE SAFETY OF EXISTING TAILINGS DAMS

Depending on the type of tailings dam and on local conditions, the stability of an existing dam can be improved by one of the following methods:

Flattening of Downstream Slope or Adding of Berms

The simplest method of improvement is achieved by the overall flattening of the downstream slope from the existing one to a slope AB as illustrated schematically in Fig. 29. Where space is available, the solution shown in Fig. 29c is more desirable because it does not require excavation into a dam which might not permit any digging, and also because a stronger type of fill can be added in compacted layers.

An example is shown in Fig. 30a. A 60 ft. high dike retaining fly ash was to be increased in two steps to 100 ft. While the starter dike consisted of a mixture of clay and partially weathered shale and sandstone, the following dike was built of fly ash with 6 in. of topsoil covering the downstream slope to develop vegetation. Since such an impervious slope cover

Fig. 27 —Sluicing with spoon for construction of tailings dam for Great Canadian Oil Sands (courtesy R. M. Hardy)

Fig. 28 —Addition of berms to existing flyash dams

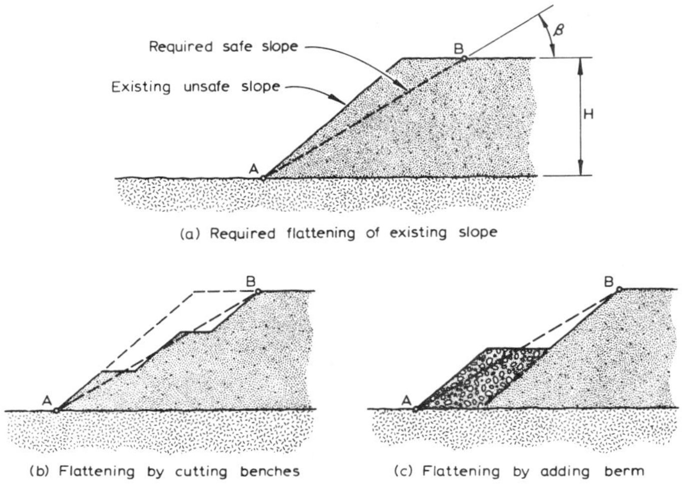

Fig. 29 —Methods of flattening existing slope

invariably causes the seepage line to rise to a horizontal position, this contributed to severe frost action and subsequent sloughing along the slope. Although the owners realized that additional dikes should not be built of fly ash, there was little doubt that failure of this dam was only a matter of time and the proposed increase in height would have hastened the occurrence. Fortunately, an unused strip of land was available between dike and highway. It permitted construction of a berm of compacted coarse granular material as shown in Fig. 30b. The subsequent addition of another 40 ft. in height offered no further problems.

Had this problem arisen without the availability of space between dike and highway, it would have been necessary to either (1) permanently detour the highway and then proceed as illustrated in Fig. 30b, or (2) dredge the fly ash from behind the dike to a depth which would have permitted replacement of the second dike by a stronger material. The second alternative would have required deposition of bottom ash, which is appreciably coarser than fly ash, and compaction along the beach as shown in Fig. 30c.

Decrease in Height of Tailings Dam

Where there is no space available for a berm adjacent to the downstream slope, and if the shell of the dam is too thin to permit cutting benches, then a decrease in height of the tailings pond may have to be considered. If space is available for transfer of tailings from the pond to another area, this is a feasible though costly solution.

Stabilization by Electro-Osmotic Tie-Backs

When it is not possible either to reduce the height of an unsafe tailings dam or to flatten its slope, there might appear to be no remaining alternatives. However, it is believed that the stability of an existing tailings dam could be increased by the installation of electro-osmotic tie-backs.

Electro-osmosis is a method of increasing the strength of soft soils by applying direct current to electrodes which are installed at a certain spacing in the soft soil to be treated (17). As illustrated in Fig. 31, two pipes or rods serving as electrodes will cause the pore water to

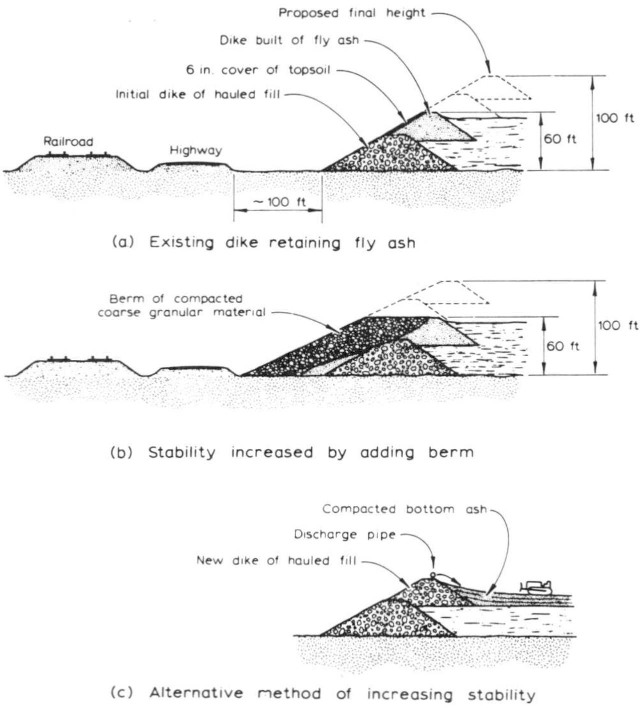

Fig. 30 —Methods of increasing stability of existing fly ash dam

move away from one electrode and collect at the other. This, together with base exchange, cementation due to deposition of metal derivates, and deposition of carbonates at one or the other of the electrodes, will quickly lead to an increased pull-out resistance of such electrodes.

Although this method has not yet been used for the purpose of developing effective tie-backs for tailings dams, it is believed that such tie-backs could improve the stability of mine tailings, fly ash, or chemical wastes. Fig. 32 illustrates how such a method could be applied to a 100 ft. high dam. The most economical spacing of the electrodes, the voltage to be applied, and the duration of treatment would vary with each case, but could be determined by simple laboratory tests.

A 3-in. diam. steel pipe imbedded to a depth of 12 ft. in very soft sandy clay developed a 10 in. diam. cemented body of soil within 10 days of electrical treatment as shown in Fig. 33. A hammer was required to enable inspection of the steel pipe. Before treatment, one man was able to push the pipe down without special effort, but it required a force of approximately 2 tons to pull the pipe up after application of electro-osmosis.

Fig. 34 shows the results of a laboratory test with a thin steel rod being treated by electro-osmosis in a very soft clayey silt. The force necessary to pull this rod increased from a negligible amount before the treatment to approximately 1300 lb/sq. ft. after two weeks of electrical treatment.

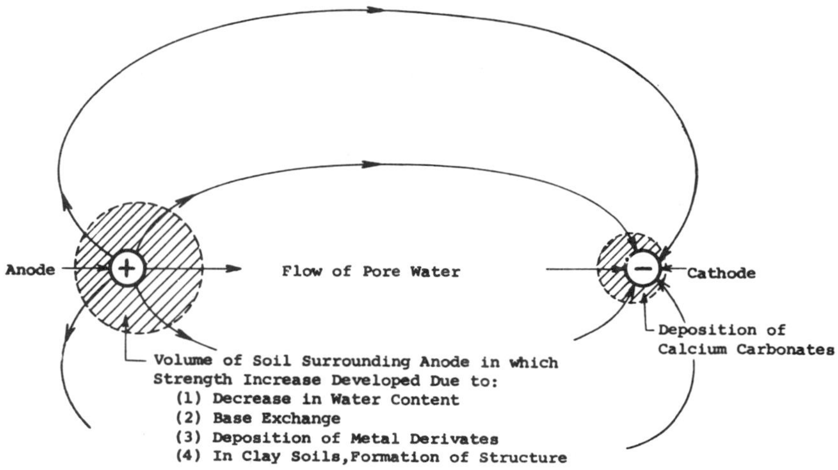

Fig. 31 —Basic effects of electro-osmosis in fine-grained materials

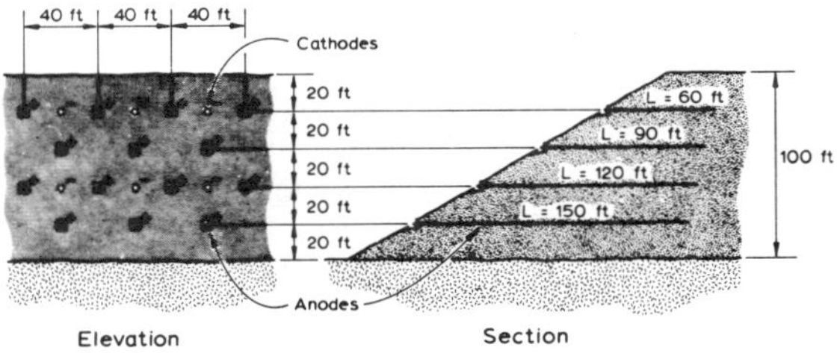

Fig. 32 —Suggested arrangement of electro-osmotic tie-backs for improving stability of tailings dam

Fig. 33 —Typical hardening of soil by electro-osmosis around 3-inch diameter pipe

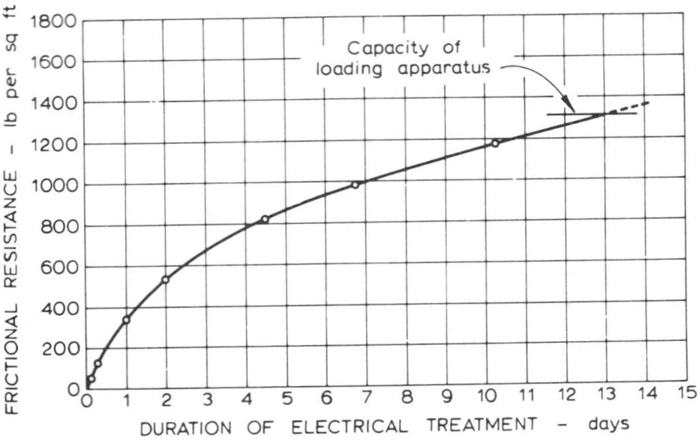

Fig. 34 —Increase in pull-out force of thin rod with time of electrical treatment

Fig. 35 —Typical spoil from strip mining

RETAINING OF SPOIL DUMPS

The topic of tailings dams would not be complete without at least a brief discussion of some of the problems involved in the retaining of mine wastes other than tailings. To dump mine spoil in a more or less uncontrolled manner, such as shown in Fig. 35 for a strip mine operation, is being increasingly frowned upon by state governments and local agencies. Apart from environmental and ecological considerations, such dumps can result in serious sliding or even liquefaction with unforeseen consequences (18). If mine spoil becomes saturated with rain or spring water discharging from the slope, this will have the following effects on loose material:

1. The unit weight of the mine spoil increases appreciably, causing the spoil to move further downhill and to adopt a flatter slope.

2. Even local readjustments of the type described under (1) above may result in the transfer of stresses into the pore water which may trigger a liquefaction failure.

3. Mine spoil containing clay will gradually become susceptible to sliding.

In the early stages of a dump, shown in Fig. 36a, many types of mine spoil have water contents which would permit excellent compaction. If the lower end of such a dump were spread and compacted in layers, as illustrated in Fig. 36b, a dam capable of resisting the instability of the spoil above would develop. With growing volume of the dump, such a dam can be increased in height and width as the need arises. Since in most instances no borrow material would be required, the cost of such retaining dams would result solely from the part-time service of a heavy crawler-type tractor and its operator.

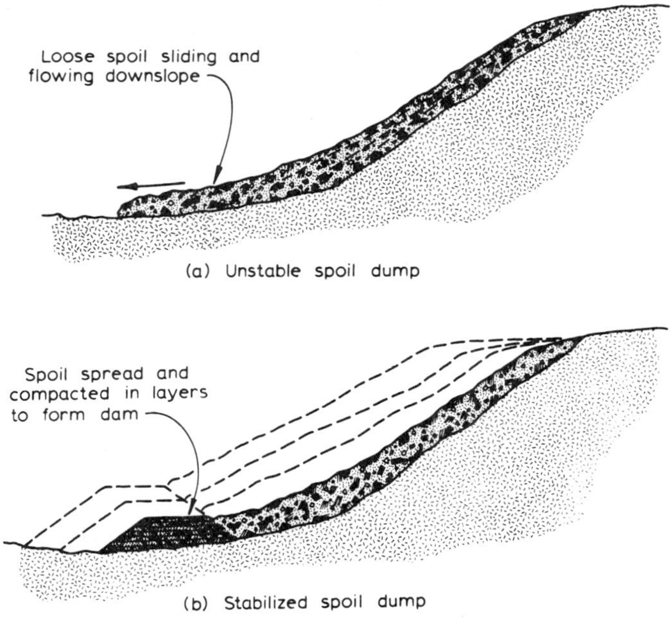

Fig. 36 —Stabilizing of spoil dump by compacting spoil at toe

CONCLUSIONS

1. Because sluiced mine tailings, fly ash, and chemical wastes are sensitive to liquefaction, considerations in the design and construction of retaining dams are similar for storage ponds of all such materials.

2. The practice of increasing the height of tailings dams by a series of dikes, each supported in part by the underlying dike and in part on soft or loose tailings, does not

assure adequate safety. This construction practice can provide safe dams only when the sand fractions of the effluent are separated and compacted at least to the axis of the final dike of the dam.

3. Fine tailings are as a rule difficult to compact and are susceptible to severe erosion and frost action; therefore, they should not be used for the construction of tailings dikes or dams.

4. Where mine spoil or other suitable materials are available close to a proposed tailings dam, consideration should be given to construction of the entire dam of compacted soil rather than only a starter dike. Such a dam can be built in stages by adding the necessary increase in height and width on the downstream side.

5. The stability of existing tailings dams that are not adequately safe can be improved either by flattening the overall slope, by decreasing the height of the pond, or by installation of electro-osmotic tie-backs.

6. Dumps of mine spoil can be safeguarded against sliding or flowing by compacting the mine spoil in layers over an adequate width along the toe of the dump.

REFERENCES

1. Casagrande, A., "Characteristics of Cohesionless Soils Affecting the Stability of Slopes and Earth Fills," Journal of the Boston Society of Civil Engineers, Jan., 1936.
2. Casagrande, A., "Notes on the Design of Earth Dams," Journal of the Boston Society of Civil Engineers, Oct., 1950.
3. Dobrey, R., and Alvarez, L., "Seismic Failures of Chilean Tailings Dams," Journal of the Soil Mechanics and Foundations Division, ASCE, Vol. 93, No. SM6, Nov., 1967, pp. 237-260.
4. Seed, H. B., "Landslides During Earthquakes Due to Soil Liquefaction," Journal of the Soil Mechanics and Foundation Division, ASCE, Vol. 94, No. SM5, Sept., 1968, pp. 1055-1122.
5. Smith, E.S., "Tailings Disposal and Liquefaction," Transactions, Society of Mining Engineering, 1968 (Rocky Mountain Minerals Conference, Las Vegas, Nevada, 1967).
6. Sherard, J.L., "Influence of Soil Properties and Construction Methods on the Performance of Homogeneous Earth Dams," U.S. Bureau of Reclamation Technical Memo No. 645, Denver, Colo., 1953.
7. Covarrubias, S. W., "Cracking of Earth and Rockfill Dams," Harvard Soil Mechanics Series, No. 82, Harvard University, Cambridge, Mass., 1969.
8. Sherard, J. L. et al, Earth and Earth-Rock Dams, John Wiley and Sons, New York, 1963.
9. ASCE Committee on Earth and Rockfill Dams, "Problems in Design and Construction of Earth and Rockfill Dams," Journal of the Soil Mechanics and Foundation Division, ASCE, Vol. 93, SM3, 1967, pp. 129-136.
10. Blight, G.E., "Shear Stability of Dumps and Dams of Gold Mining Waste," Transactions, South African Institution of Civil Engineers, Vol. 11, March, 1969.
11. MacIver, B.N., "How the Soils Engineer can Help the Mill Men in the Construction of Proper Tailings Dams," Engineering and Mining Journal, May, 1961.

12. Hardy, R.M., "Earth Dams in Western Canda," Western Congress of Engineering, May, 1970.

13. Windolph, F., "Tailing Pond Design," Mining Engineering, Vol 13, No. 11, 1961.

14. O'Connell, C.A., "The Mufulira Western Expansion Project Increases Production 50%", Mining Engineering, Vol. 14, No. 12, 1962.

15. Burke, H.H., "Design and Construction of Dams for Retention of Solid Wastes," ASCE National Meeting on Water Resources Engineering, New Orleans, Louisiana, Feb., 1969.

16. Kealy, C.D., and Soderberg, R.L., "Design of Dams for Mill Tailings," U.S. Dept. of the Interior, Bureau of Mines, Information Circular 8410, U.S. Government Printing Office, Washington, D.C., 1969.

17. Casagrande, L., "Electro-Osmotic Stabilization of Soils," Journal of the Boston Society of Civil Engineers, Jan., 1952.

18. National Coal Board (Welsh Office), "A Selection of Technical Reports Submitted to the Aberfan Tribunal" (3 Volumes), Her Majesty's Stationary Office, London, 1969.

Chapter 12

CASE STUDIES OF STABILITY ON MINING PROJECTS

by

C. O. Brawner

Principal, Golder, Brawner and Associates, Vancouver.

INTRODUCTION

The procedures used to design slopes in rock have only been developed over the past 10 years. Therefore, continued review and assessment of field experience by the publication of case studies is necessary to determine the accuracy of the theory and the application of the design techniques being used.

Where failures have occurred, mine management should insist that the slides be 'back analyzed' to determine the shear strength parameters at failure. These values should be used for any pit design as they reflect site conditions more accurately than any other procedure.

The previous papers have dealt with the basic concepts and factors that influence stability. To illustrate the application of these concepts in the study of practical stability problems, six case studies have been selected for review. The examples include failure of tailings dams due to an earthquake, failure and redesign of a slope in overburden, review of failures of rock slopes, calculation of the amount of seepage into a pit and influence of the groundwater on stability, and the design of slopes for a proposed 2,500 ft. deep open pit mine. All but the first example are drawn from the authors personal experience.

EXAMPLE 1

Failure of Tailings Dams Due to an Earthquake

On March 28, 1965 a serious earthquake occurred in Chile with the epicenter located near the town of La Liqua, about 80 miles north of Santiago. The earthquake had a rating of 7 to 7.25 on the Richter scale. A record of the acceleration was taken at Santiago where a maximum horizontal acceleration of 0.18 g. was recorded.

Many small mines are located in this region. Each of the mines provided impoundment storage for tailings with one to three dykes. As a result of the earthquake, most of the dams failed. The most serious failure involved two dams at El Cobre where more than 2,700,000 cu. yd. of tailings flowed into the valley, travelled over 7 miles in a few minutes, destroyed part of the town of El Cobre and killed more than 200 people.

The Chilean government commissioned an investigation into the failures. A summary of the findings was published by Dobry and Alvarez in 1967 (1).

The location of the earthquake and the tailings dams which were investigated is shown in Fig. 1.

Most of the dams were constructed as hydraulic fill structures. The tailings were deposited on the top of the outer shell through pipes or flumes so that the coarsest sands were deposited on or near the dyke, and the finer portion or slimes flowed into the pond area. Studies indicated that the percentage of particles smaller than 0.005 mm. increased gradually up to about 100 ft. from the dyke and then remained constant beyond that point.

A total of 22 tailings dams in 10 areas were studied. These were located between 11 and 60 miles from the epicenter of the earthquake. The height of the dams ranged from 15 to 100 ft. The exterior slopes varied from 15° to 40° and most averaged 33°. The majority of the dams were built on natural ground which sloped less than 7°. The dykes were generally constructed to surround the impoundment area comprising a square, rectangular or triangular shape.

The failures involved cracking of the dykes, failure of the exterior slopes of the dykes and release of fluid tailings that flowed up to several miles from the impoundment area. At 6 locations, the volume of tailings that liquefied and flowed exceeded 30,000 tons. In 8 of 9 locations where detailed soils tests were performed, the natural moisture content of the tailings

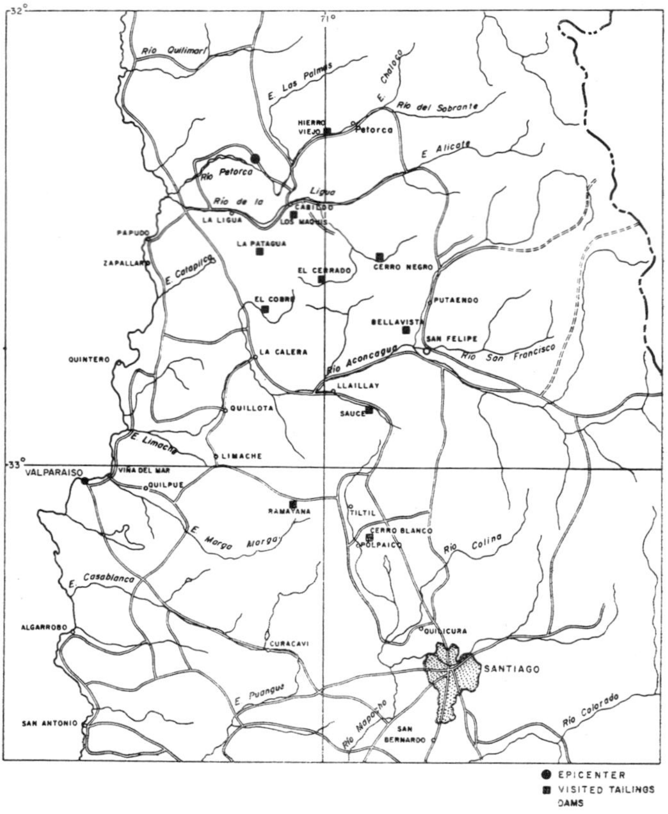

Fig. 1 — Map showing location of earthquake and tailings dams. (1)

was at or above the liquid limit. Gradation tests on the shell materials revealed that 30% to 61.5% was finer than the No. 200 screen. The average gradation of the tailings deposited at El Cobre between January and October, 1964 is shown on Fig. 2. Standard penetration tests, to assess relative density, were performed at El Cobre. Below the upper layer, which was dried out due to desiccation, the tailings were described as unconsolidated with zero penetration resistance and the natural moisture content was greater than the liquid limit.

Table I summarizes details of 11 dykes that failed. The metal mined at all but the Cerra Blanco mine is copper.

The largest and most serious failure occurred at El Cobre. Three dams had been developed, two of which were commenced in 1930 and were constructed using gravity separation from flumes. Of these two, one dam was out of use and the other was used only for emergency disposal. A new dam was constructed in 1963 with a shell constructed from the coarser fraction that had been separated by cyclones. The angle of the exterior slope of the new dam was only 15°. Details of the El Cobre dams are shown in Table II.

Eyewitnesses who observed the failure of the new dam indicated that the earthquake initially produced waves on the surface of the ponded tailings. This was followed by a crack that opened at one corner of the triangular shaped dam. The liquefied tailings started to flow through this crack, rapidly eroding the opening allowing about 500,000 cu. yd. of tailings to flow out.

The investigators postulated that the earthquake liquefied the tailings behind the shell, and that the pressure of the liquefied material on the exterior embankment, added to the horizontal forces of inertia, caused the shell to crack and release the tailings.

At the larger of the two old dams, the front slope of the south corner failed and over 2,000,000 cu. yd. of unconsolidated tailings flowed out. The profile of the old dam before and after the failure is shown on Fig. 3.

These failures illustrate the susceptibility of loose, fine grained, saturated soils to liquefication as a result of seismic forces. In recent years, many dams have used cycloned tailings for shell construction. As a result of the failure of the new dam at El Cobre, it is apparent that tailings dams constructed with the coarser cycloned fraction deposited by gravity at moderately flat slopes are also susceptible to failure unless they are adequately compacted. It is also obvious that the corners of dams are critical areas for stability.

The investigation concluded that it is possible to design and construct tailings dams capable of withstanding earthquake forces by following the standard procedures used for the construction of earth dams. It was recommended that every future design of this type include at least the following:

(a) A clear separation between the resistant embankment and the non-consolidated core, with special precautions taken to prevent the formation of weak strata or zones.

(b) Compaction of the shell.

(c) Adequate drainage.

The most important of these requirements is adequate compaction. Good engineering practice also requires the following conditions be satisfied.

(a) The foundation soils under the shell must be sufficiently strong to provide adequate bearing capacity.

(b) Drainage under the dam must be controlled to minimize release of chemically charged tailings water, and to resist piping.

(c) The dam must be designed so that flood water can never overtop the dam.

(d) Culvert or pipe installations through the dam must be designed to prevent piping around the conduit.

Locality	El Cobre	El Cobre	Hierro Viejo	Los Maquis	La Patagua	Cerro Negro	El Cerrado	Bellavista	El Sauce	Ramayana	Cerro Blanco
Distance to epicenter, in kilometers	40	40	18	15	22	38	37	55	66	85	96
Name of dam	Old	New	—	No. 3	New	No. 3	—	—	New	No. 1	—
Height, in meters	32-35	15	5	15	15	20	25	20	6	5	8-10
Maximum exterior slopes, in degrees	35-40	15	35-40	30-35	35	35-45	35	30-35	30-35	33	34
Daily accumulation, in tons	0[c]	2,200	23-30	40-50	190	265	0[c]	80	35	12-15	—
Percentage water in volume of deposited tailings[a]	85	85	—	90	70	—	—	70	70	65	60
Water recovered, in percentage	60	—	—	70	80-85	—	—	—	—	30-40	—
Slope of foundation soil, in degrees	4	4	1-2	15-20	3-5	20	10-12	7	5	30	1-2
Permeability of foundation soil, in centimeters per second	10^{-8}	10^{-8}	low[d]	medium[d]	medium to high[d]	high[d]	medium[d]	medium[d]	low[d]	low[d]	medium[d]
Shape of scarp and/or fractures	Semi-circular scarp	Semi-circular scarp	Irregular scarp	Semi-circular scarp	Semi-circular scarp	Semi-circular scarp	Semi-circular scarp	Semi-circular scarp	Fractures parallel to border	Semi-circular scarp	Fractures parallel to border
Flowed material, in tons	1,900,000	500,000	1,200	30,000	50,000	120,000	—	100,000	—	200	—
Distance reached, in kilometers	12	12	1	5	5	5	—	2.5	—	foot of mountain	—
Core sample											
Percentage smaller than mesh No. 200	92.8	90	99.6	100.0	99.8	100.0	—	87.5	92.9	99.5	—
Liquid limit, in percentage	19-47.7	26.6	54.7	35.1	42.8	47.0	—	25.6	28.5	48.7	—
Plasticity index	0-19	4.6	30.5	8.7	17.8	17.5	—	3.4	4.7	17.9	—
w_{nat}/LL	1.18	1.00-1.50	0.83	1.11	0.99	0.99	—	0.98	1.35	1.14	—
Shell samples											
Percentage smaller than mesh #40	95	97	98.8	99.9[e]	99.8	100.0	89.0	98.6	99.8	94.5	100.0
Percentage smaller than mesh #60	75	80	93.8	96.9	97.1	99.1	74.5	83.1	98.0	71.4	99.9
Percentage smaller than mesh #200	45	30	61.6	46.3	55.3	51.6	42.9	43.4	48.7	25.4	94.7
Liquid limit, in percentage[b]	18	19	19.6	17.8	21.3	22.8	17.2	17.1	21.3	21.6	30.9
Natural moisture content, in percentage	15[e]	15	15.4	6.0	13.9	12.2	1.5	1.7	10.3	7.8	9.37
Observations	Gravitational method	Cyclones	Gravitational method	Gravitational method	Gravitational method	Gravitational method	—	Gravitational method	Gravitational method	Gravitational method	Riprap wall

[a] Proportion of total volume
[b] The shell samples had no plasticity
[c] Abandoned or emergency dam
[d] Estimated
[e] The sampling was done in another similar dam which did not fail.

Table 1 — Summary of details of dams and failures (1)

Name	Construction Method	Dimensions (Ft.)	Exterior Slope	Volume (cu. yd.)	Slide Details (cu. yd.)
Old dam (not in use)	Hydraulic	1200 x 1200 100 ft. high	35°-40°	5,000,000	2,200,000 flowed
Small dam (emergency use only)	Hydraulic	400 x 725 80 ft. high	35°-40°	1,150,000	Only local slides
New dam	Cyclones	Triangle 640 (60 ft. high)	15°	585,000	585,000 flowed

Table 2 — Details of El Cobre dams (1)

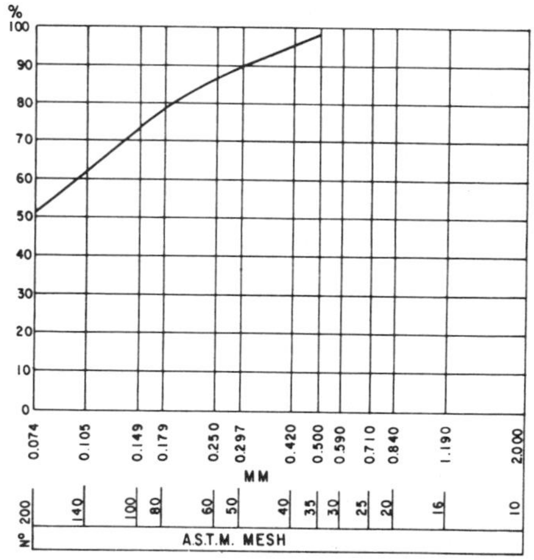

Fig. 2 —Average gradation of tailings deposited at El Cobre between January and October 1964. (1)

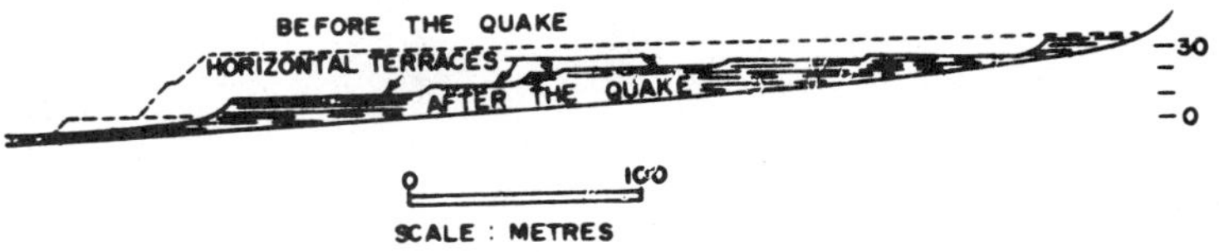

Fig. 3 —Profile of the old dam before and after failure. (1)

EXAMPLE 2

Stability of a Slope in Deep Overburden Soil

In the north east area of a major open pit mining project, overburden comprising stratified granular and cohesive fine grained soils up to 300 ft. in depth overlies the ore body. These slopes had been excavated to an angle of 34° to 37°.

In May 1968, a tension crack opened beyond the top of the slope. Since the mining operation for 1968 included removal of overburden, it was possible to remove about 50 ft. of the upper slide area and stabilize the movement.

In the spring of 1969, a tension crack was observed behind the top of the slope and the slope started to move noticeably on June 2. At the same time, the lower portion of the overburden face, which coincided with a main haul road, began to move outward. Fig. 4 shows a photograph of the slope. Considerable water was observed seeping from the toe area. In order to maintain the haul road, the slide material that encroached on the road was excavated periodically. The major movement occured in the late stages of spring, which probably coincided with a high water table condition.

An investigation was commenced to determine the soil and groundwater conditions, to attempt to learn the cause of the slide, to develop a stabilization program and to establish design criteria for new slopes that would be developed as further overburden was removed.

The field program comprised drilling five boreholes through the overburden into the underlying bedrock. Samples were obtained of each soil type, standard penetration tests were performed to assess relative density and falling head permeability tests were carried out to measure permeability. The surface of the slope adjacent to each borehole was mapped and the exposed soils were sampled. Disturbed and undisturbed samples were obtained during drilling and chunk samples were cut from the slope for laboratory tests. After each borehole was completed, piezometers were installed to measure the hydrostatic pressure. Fig. 5 shows the outline

Fig. 4 —Photograph of the face of the overburden slope which was moving.

of the slide area and the locations of the boreholes. Fig. 6 shows a generalized stratigraphic section through borehole 4 in the slide area.

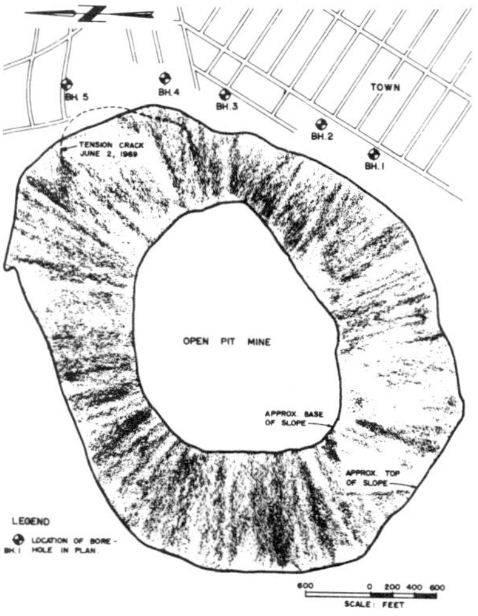

Fig. 5 —Location of overburden slide and boreholes drilled for investigation.

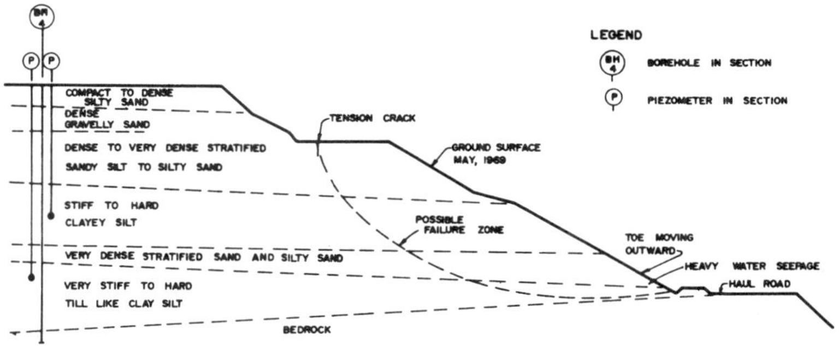

Fig. 6 —Generalized section through borehole 4 in the slide area.

Four general soil types were observed in the slide area. There was a dense to very dense stratified sandy silt, a dense stratified silty sand, a stiff to hard clayey silt and a very stiff to hard till-like clay silt.

The piezometers revealed two water tables and considerable seepage was observed issuing from the face of the slope. Laboratory tests were performed on each soil type to measure the effective stress shear strength parameters. Tests included the consolidated drained direct shear test, consolidated undrained triaxial compression test and the consolidated undrained triaxial compression test with pore pressure measurements.

Stability analyses using the Bishop Modified method were performed on the computer. The initial study utilized the 'back analysis' technique. Using a factor of safety of 1.0 and hydrostatic pressures determined in the field program, a section through the centre of the slide was analyzed to determine the effective stress shear parameters. Based on this analysis and a study of the laboratory shear strength data, the following parameters were selected for stabilization analysis and slope redesign.

Soil Type	**ϕ' degrees**	**c' p.s.f.**
Stratified sandy silt	36	0
Stratified silty sand	38	0
Clay silt	30	200
Till-like clayey silt	30	300

Stability analyses revealed that a lower safety factor existed for the lower portion of the slope than for the overall slope. This confirmed field evidence that the slide had developed retrogressively with the lower slope failing first, relieving support for the toe and resulting in a backward extension of the slide.

Analyses were also performed to determine the influence of groundwater on stability. By reducing the "r_u" factor, ratio of $\left(\frac{\text{pore water pressure}}{\text{total pressure}}\right)$ from 0.2 to 0.1 the safety factor was theoretically increased by 20%. If the water pressure is reduced this amount, the slope can be steepened about 4° while maintaining the original safety factor.

Horizontal drains have proven effective in stabilizing many slides (2) (3). In order to determine whether the clayey silt soil in the slope could be drained, a test program employing drains was used. Five drains, about 120 ft. long, were installed. The drains consisted of 2 in. dia. pipe with ⅜ in. dia. drain holes spaced 3 in. apart in two rows along the pipe. They were

Fig. 7 —Photograph of horizontal drain installations. The left drain is flowing 3,000 gallons per day and the right drain 19,000 gallons per day.

installed in holes drilled at 5° above the horizontal and fanned out from one drill set up. Fig. 7 shows two of the drains flowing water. Piezometers were installed in two vertical drillholes, prior to drain installation, to determine if the water table would be lowered by the drains. Following completion of the test installation, the piezometric levels dropped 15 ft. in 20 days. This indicated that horizontal drains would be effective in lowering the water level.

The data developed from the 'back analysis' of the slope failure were used for design of future slopes in the overburden soil. Design charts were prepared for different groundwater conditions and for the variable soil profiles that the drilling indicated exists around the pit. In addition, the design charts were prepared for various safety factors. The higher the safety factor, the lower the probability of slope failure. Fig. 8 is a typical design chart prepared for the area of the slope near the failure.

It is normal practice in soil mechanics, where long term stability is required, to use a safety factor of 1.5. On mining projects, however, overburden slopes are frequently exposed for short periods. Also, since overburden represents waste excavation, there is a great incentive to utilize slopes as steep as possible to reduce excavation costs. If the safety factor is reduced to near 1.0, safety considerations require that the slope be continuously monitored to pick up early indications of impending movement. This will usually allow time for stabilization.

Where the cost of overburden removal on a mining project is great, design refinement should be considered. This could include the use of a variable slope, strengthening the toe area of the slope where shear stress is highest, or the use of special drainage techniques if pore water pressures exist in the slope.

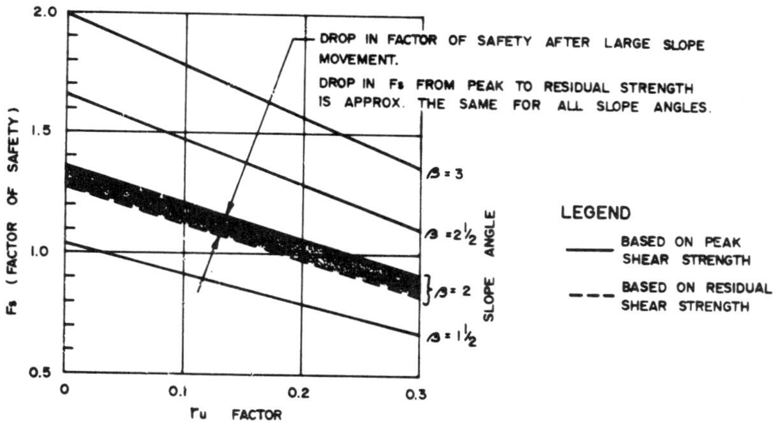

Fig. 8 —Design chart for determination of slope angle for various safety factors and pore pressures.

EXAMPLE 3

Pit Slope Failure in Rock

On April 15, 1969 several sink holes were observed in a drainage ditch approximately 400 ft. behind the crest of a 450 ft. high open pit face. On the following day, cracks were observed to have developed from 50 to 150 ft. behind the crest of the slope, and on April 18th a slide occurred. The crest dropped an average of 70 ft., noticeable bulging occurred in the body of the slope and extensive rock falls developed on the face. Initially it appeared that the toe had not moved, however subsequent checking of cross sections before and after the slide indicated that up to 20 ft. of movement had occurred. The total volume of material involved in the slide approximated 800,000 cu. yd. Fig. 9 shows a photograph of the slide looking from the end of the pit and Fig. 10 shows a plan of the slide area.

Two major shear zones were exposed at the back of the slide, along which some of the

Fig. 9 .—Failure of rock slope. Note steeply dipping shear zones at the top of the slide.

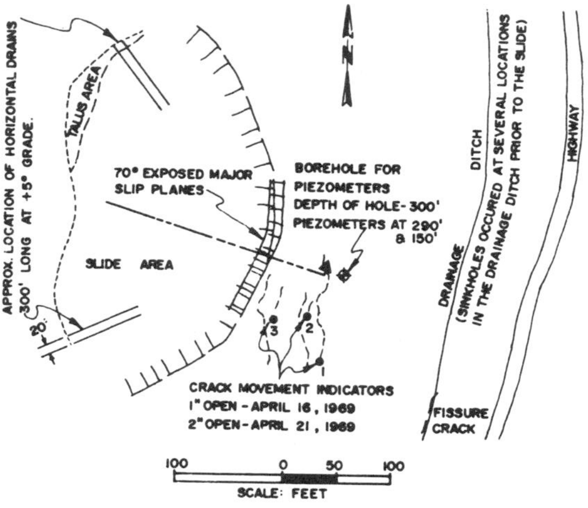

Fig. 10 —Plan of slide area.

movement obviously occurred. One shear zone is oriented approximately north 20° west and the second shear zone at south 30° west. Both shear zones dip at an average of 70° below the horizontal. These shears appear to be part of a regional system of faults since several other such discontinuities occur with similar orientation along the same side of the pit. The slope that had failed consisted of a portion of a promontory that projected out from the slope into the pit. The front slope angle of this promontory was about 40°, whereas the remainder of that side of the pit had been excavated with an average slope of 45°. Because of the large number of precarious rock falls covering much of the failed slope, it was not possible to determine whether moderately flat lying joints or shear zones existed in the toe area of the slide. However the mechanism of the movement suggests that this may be the case.

The rock type in the failure area is predominantly serpentine.

Shortly after the failure occurred, movement hubs were installed on both sides of the cracks that had developed behind the slide. After a period of 48 hours, 1 in. of further movement had been observed.

The control of drainage beyond the slide is of considerable significance. A major highway is located about 600 ft. beyond the top of the slide and a railway is located approximately 700 ft. away. Drainage, which is carried in culverts under these two structures, is intercepted by the collector ditch located between 50 and 400 ft. behind the top of the pit backslope. This ditch carries the water around the periphery of the pit and deposits it in natural drainage channels. It was in this ditch that the sink holes were observed prior to the slide. Considerable seepage water issues from the face of the slope below the ditch, particularly following periods of heavy rainfall or during spring breakup.

Average precipitation in the area of the pit is approximately 40 in. per year.

Several factors are believed to have contributed to the movement. The more important of these are as follows:

(a) Excess cleft water pressure. This slide occurred during the spring breakup period when the drainage ditch beyond the backslope was carrying considerable water. As a result, the water table in the slope would be expected to be at a high level. In addition, any surface cracks that opened would allow the infiltration of water and the development of hydrostatic forces. There was a lack of snow on this side of the pit during the winter so that a considerable depth of frost penetration could have occurred. It is possible that this face may not have thawed out prior to the slide. This could have restricted the flow of water from the slope so that abnormally high cleft water pressures could have built up.

(b) Rock Structure. It is obvious that the two major shear zones, which dip at about 70° out of the slope, had a major influence on the instability. It is suspected that a system of moderately flat lying discontinuities dips out of the slope, along which failure at the toe occurred. However, because of the cover of fallen rock, it has not been possible to definitely establish this condition.

(c) Rock Strength. There is a possibility that the presence of water in the slope may have reduced the coefficient of friction in the rock. This reduction is unusual and does not occur in massive structured minerals such as quartz. However, Horn and Deere (4), have performed tests on Vermont serpentine that showed a reduction in the tangent of the friction angle from 0.50 to 0.29 from the dry to saturated state.

(d) Pit Slope Geometry. The area that failed consists of a promontory of rock in the east face of the pit. It has been shown theoretically by Jenkie and Yen (5) that convex shaped slopes are less stable than straight faced slopes or concave slopes, all other factors being equal. The main reasons for this are a reduction in both the side restraint and area of the failure surface.

(e) Increase in Seismic Stresses. During the months prior to the failure, the size of blasts in the pit were gradually being increased and the number reduced from one blast per day to two blasts per week. It is possible that the seismic acceleration forces and potential hydrodynamic shock in the rock may have increased as a result of the larger blasts (6).

Surface movement hubs were installed on top of the slope on both sides of the cracks. These were monitored daily for a period of several weeks after the slide. In addition, one borehole was drilled approximately 125 ft. behind the shear face of the slide. This hole was drilled to a depth of 300 ft., and piezometers were installed at depths of 150 and 290 ft. Details of the installation are shown on Fig. 11. The water level was determined to be 104 ft. below ground surface in the shallow piezometer and 133 ft. below the surface in the deeper piezometer. This indicates the presence of a discontinuous water table.

In view of the high water level within the slope, the existence of the drainage ditch near the top of the slope, and visual evidence of considerable seepage from the face of the slope, it was obvious that water pressures contributed significantly to the reduction in stability of the slope.

Stability analyses were performed on a cross section at the centerline of the slide. For a reduction in the level of the water table of 50 ft. in the slope, the factor of safety was increased by 35%. Because of the significant effect of water on stability, and the difficulty of obtaining access to the top of the slope to remove material, which would also improve stability, recommendations to control further movement were directed towards reducing the cleft water pressures. The recommendations to improve stability included water proofing the drainage ditch just behind the top of the rock slope in conjunction with the diversion of any of this water that was practical, filling all open cracks behind the slide to reduce seepage into the slope, and the installation of horizontal drains at the toe of the slope.

To determine whether the drains would be successful, test installations, consisting of two drains in the north side of the slide and two drains on the south side of the slide, were installed to a depth of approximately 300 ft. No water flowed from the drains on the south side of the slide. However, the drains on the north side produced up to 1600 gph. (Fig. 12). After completion of the first drain in this area, the water level in the piezometers commenced to drop rapidly (Fig. 13). A reduction in cleft water pressures of 30 to 40 ft. occurred in the piezometers within 30 days after the drains were installed.

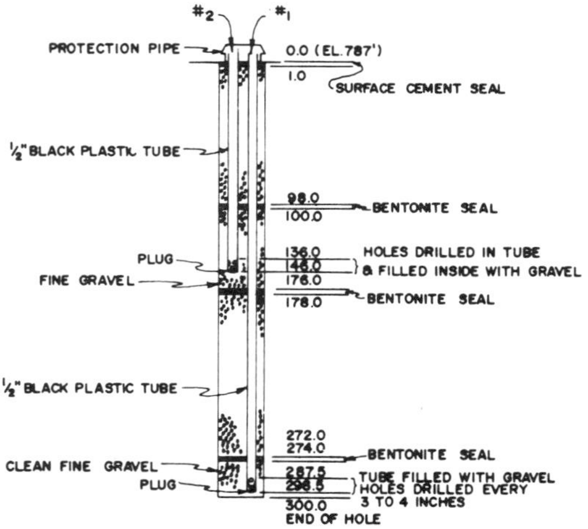

Fig. 11 — Details of piezometer installation.

From this program, it was inferred that a system of horizontal drains along the toe area of the slide and along the toe area of the remainder of the east slope would provide a significant benefit in improving the overall stability.

This example also reveals the importance of structural discontinuities and pit geometry on stability.

Fig. 12 — Horizontal drain at toe of slope flowing 1,600 gallons per hour.

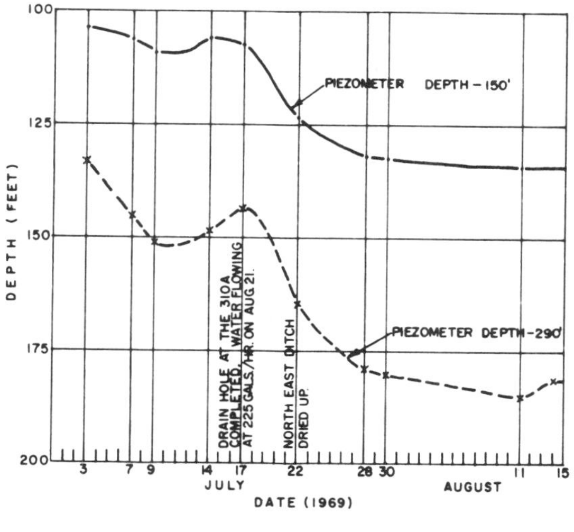

Fig. 13 — Piezometer levels following slide and installation of horizontal drains.

EXAMPLE 4

Pit Slope Failure at an Open Pit Coal Mine

At an open pit coal mining operation, up to several hundred feet of shale and sandstone is being removed to expose a coal seam approximately 40 ft. thick.

The original ground surface dipped to the west at 15 to 25°. The underlying sedimentary structure dips approximately 30° to the west near the top of the ridge, gradually flattening towards the valley floor, which is approximately 800 ft. lower in elevation. Exploratory drilling indicated that the coal is underlain by basal sandstone containing occasional shale layers.

As the mining program developed, and as excavation reached approximately 200 ft. below the surface exposure of the coal seam, a slide occurred in the backslope (Fig. 14). This was evidenced by a drop of about 6 ft. at the top of the pit slope and development of an open tension crack that was exposed for about 500 ft. along the crest of the slope. The toe of the failure was observed as a zone of push up in the slope, approximately 200 ft. below the top of the slope. The volume of material in the movement was estimated at 50,000 cu. yd.

An investigation was performed to determine the sub-surface conditions, to determine the probable cause or causes of failure, to develop procedures for stabilization, and to evaluate stability as the height of the backslope increases with further mining.

Three boreholes were put down with a diamond drill. Two were located on the lowest working bench that had been developed, and one was put down on the approximate centerline of the slide, about 75 ft. below the upper crack. Continuous core was recovered and mapped structurally, and representative samples of the rock were submitted to the laboratory fc⁻ shear tests. After completion of the drilling, water packer permeability tests were performed in the boreholes at 10 ft. intervals. This was followed by the installation of piezometers to measure cleft water pressures.

The general stratigraphy of the slope is shown in Fig. 15. In the two lower boreholes, 20 ft. of medium hard, dark grey shale was encountered, which in turn was underlain by approximately 120 ft. of interbedded dark grey shale with light grey laminated sandstone layers. The

Fig. 14 —Photograph of pit slope showing slide.

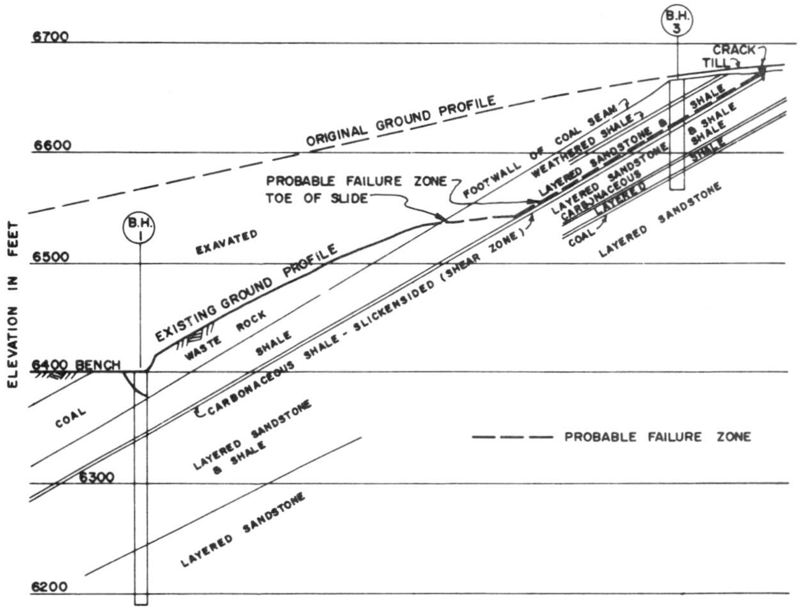

Fig. 15 —General stratigraphy and probable failure zone.

thickness of the interbedded layers varied from 0.3 to 6 ft. with a structural dip of 28° to 30°. At a depth of 43 to 45 ft., a highly fractured carbonaceous shale with a trace of clay was encountered. Slickensides were observed in the core. In the upper borehole, weathered shale was encountered to a depth of 15 ft., underlain by interbedded shale and sandstone to 87 ft., which was the depth of the hole. Fractured and slickensided zones of carbonaceous shale were encountered at depths of 16 to 18 ft., 30 to 33 ft., and 72 to 73 ft. It appears probable that the primary zone of failure occurred at a depth of about 30 to 33 ft. below the base of the coal seam in a thin layer of carbonaceous shale. This depth corresponds well with the projection of the upper crack parallel to the bedding.

The water packer permeability tests indicated a range in average permeability of 6 x 10^{-4} cm./sec. to 2 x 10^{-6} cm./sec. Maximum permeability would be in the direction of the bedding.

The water table was encountered at a depth of 162 ft. in the lower boreholes. An extended period of dry weather appears to be the reason for this low water level. Heavy rainfall occurred shortly after the installation of the piezometers and the piezometric level rose to within 36 ft. and 6 ft. of the top of the casing in boreholes 1 and 2 respectively. This rapid change in levels illustrates the response of cleft water pressures to precipitation and spring runoff. It is believed that the unloading of the slope by excavation resulted in some opening of the bedding joints, which allowed higher than normal infiltration of water into the rock at the top of the slope.

A cross section of the slide was 'back analyzed' to determine the average shear strength parameters for various cleft water pressures. The failure surface was assumed to be the slickensided carbonaceous shale layer that exists about 35 ft. below and parallel to the final pit face. Assuming no cleft water pressures in the slope, the average effective angle of friction for a safety factor of 1.0 was computed to be 30°. Assuming a piezometric pressure equivalent to 0.15 of the normal pressure on the failure plane, the average effective angle of friction for the same safety factor was computed to be 35°.

The movement occurred quite rapidly and then stopped. It is believed that high pore water pressures developed in the carbonaceous shale layer which reduced the effective stress and shear strength and caused the movement. As movement progressed, dilatation occurred and the pore pressures partially dissipated. This would increase the effective pressures and shear strength and result in the movement stopping. Future pore pressure build up could again cause movement to start.

Direct shear tests were performed on the shale. At low normal stress the angle of friction averaged 35°. This gradually decreased to 26° with an increase in the normal stress. An average angle of friction of 32° was assumed for the stress range encountered in the field.

Based on the angle of friction computed from the 'back analysis' and determined from the laboratory testing, stability studies indicated that continued extension of the 29° slope would be stable provided that piezometric pressures were not allowed to build up in the carbonaceous shale zone. To control the water pressure, horizontal drains, spaced 50 ft. apart, drilled on a plus 5% grade, and carried 15 ft. beyond the shale layer were recommended. Since the depth of the failure zone is shallow, a relatively small increase in piezometric pressure can create an unstable condition. Therefore, the drains must be installed from every bench. To provide assurance that the drains function as required, piezometers must be installed along every bench to monitor the water pressures.

EXAMPLE 5

Investigation of Seepage Volume and Influence of Groundwater on Stability

A proposed pit is located adjacent to a major inlet on the British Columbia coast. The ultimate depth of the pit will be about 600 ft. below sea level.

An investigation was performed to determine the seepage potential from the ocean inlet into the pit and to determine the influence of cleft water pressures and seepage on stability.

The topography over the pit site is undulating and contains occasional rock outcrops. The depth of overburden ranges up to 200 ft. It consists of dense to very dense, clayey, silty glacial drift with occasional sand and gravel layers.

The bedrock generally consists of massive, fine to medium grained, andesite and andesite breccia with numerous fractures, many of which are recemented with carbonates. Occasional weathered recemented fractures and gouge filled shear and fault zones were encountered.

The proposed pit is adjacent to the tidal shoreline for a length of about 4,000 ft. The crest of the pit adjacent to the shoreline is from 15 to 75 ft. above sea level and the horizontal distance from high tide level to the edge of the pit ranges from 100 to 600 ft.

A major concern is whether the salt water seepage from the inlet will find its way into the pit. It was necessary to estimate the volume of seepage that could be expected and the effect it would have on stability.

Boreholes were drilled using HQ and NQ size drill equipment. The overburden was sampled and falling head permeability tests were carried out to estimate 'in situ' permeability. Bedrock was cored using NQ coring equipment and continuous pressure packer tests in 20 ft. sections were carried out to determine the average permeability of the bedrock.

All bedrock core was logged for fractures and faults and the dip of these was determined. During a previous investigation for rock slope design, structural mapping was performed in an adit and in four boreholes that had been drilled to obtain core during proving of the ore body. These boreholes were viewed with a borehole periscope (Fig. 16). The deter-

Fig. 16 —Borehole periscope used to map rock structure in boreholes.

mination of joint frequency and structural orientation was required to assess directional permeability.

The coefficient of permeability of the overburden till stratum was determined to range from 10^{-4} to 10^{-7} cm./sec., and the coefficient of permeability of the coarse granular layers within the overburden was determined to be in the order of 10^{-2} cm./sec. The coefficient of permeability of the bedrock ranged from 10^{-5} to 10^{-9} cm./sec. with a general range of values in the order of 10^{-6} to 10^{-7} cm./sec.

Seepage studies were performed to estimate the volume of seepage water to be expected. Flow nets were drawn within the overburden and bedrock with the variable layered conditions being taken into account. Fig. 17 shows a typical flow net in the bedrock.

Based on permeability values and flow nets determined by the test program, it was computed that about 1,000,000 gal. per day could be expected as seepage from the pit face adjacent to the inlet. This volume is not critical since rainfall of 4 in. per day occurs several times a year in the area, which amounts to about 6,000,000 gal. per day over the same area, or 6 times the estimated seepage. Therefore pumping capacity must be designed primarily to handle precipitation volumes.

Two dimensional, stability analyses were performed to determine the influence of cleft water pressures and seepage pressures on pit slope stability. In the overburden, it was computed that slopes could be excavated at 35° above the water table and 20° below the point where the water table intercepted the slope. Drainage to lower the water level in the overburden will therefore significantly reduce excavation volumes. The most effective means of drainage for this slope is considered to be the installation of horizontal drains.

In the bedrock, slope stability analyses indicated the slopes can be steepened by about 5° to 7° if cleft water pressures and seepage pressures are not allowed to develop within

100 ft. of the rock face. If the salt water seepage is not detrimental to equipment or metallurgy, horizontal drains will be the most effective and economical means of drainage. However, if the salt water will be detrimental, total interception of the seepage by deep wells or underground adits may be necessary.

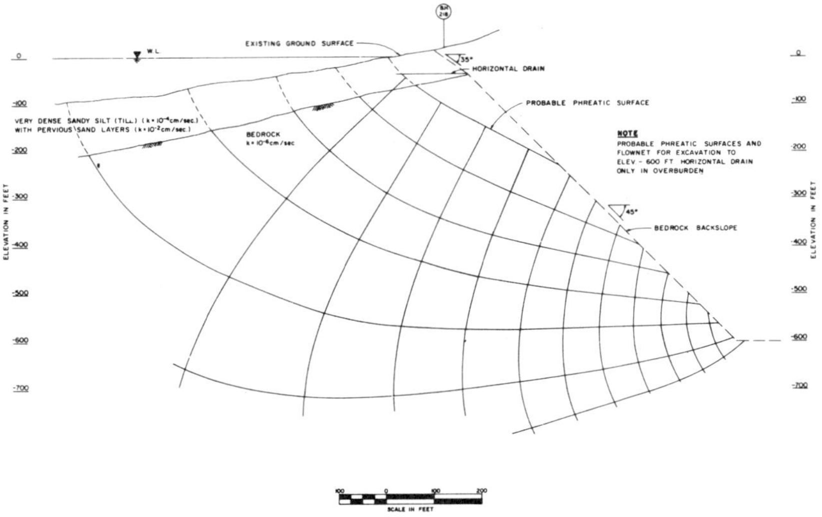

Fig. 17 —Typical computed flow net in bedrock with horizontal drains in overburden only.

EXAMPLE 6

Preliminary Design of Slopes for a Pit 2,500 ft. Deep

The proposed pit is located in the bottom of a mountain valley at an elevation of 4,000 ft. The rock is generally diorite, granodiorite and porphyry. One major fault has been indicated that runs through the proposed pit with a strike of about north 30° west and a dip of 70° to 90° west.

Slope design studies have been performed for a potential pit 2,500 ft. deep.

The orientation of the rock structure was measured by logging and mapping rock core obtained from the ore drilling program. Six of the boreholes drilled during this program were protected and later surveyed with a borehole camera. Fig. 18 shows the camera equipment and Fig. 19 shows a typical borehole camera photograph.

Structural mapping of the walls and ceiling of 5,000 ft. of exploration adit was also performed. The structural data was divided into open joints, tight joints, shears and faults. Computer techniques were used to prepare polar plots for structural analysis. A typical plot obtained from a computer plotter print out is shown on Fig. 20. In the diagram, letters of the alphabet are used for numbers of discontinuities in excess of 9.

The analysis revealed two preferred orientations; a strike of north 70° west with a dip of 70° south and a strike of north 18° west with a dip of 62° east. In addition, significant joints, fractures, and shears occur with an easterly dip, generally steeper than 40° with a strike azimuth ranging from 180° to 350°.

Fig. 18 —Borehole camera equipment used to photograph rock structure in deep boreholes.

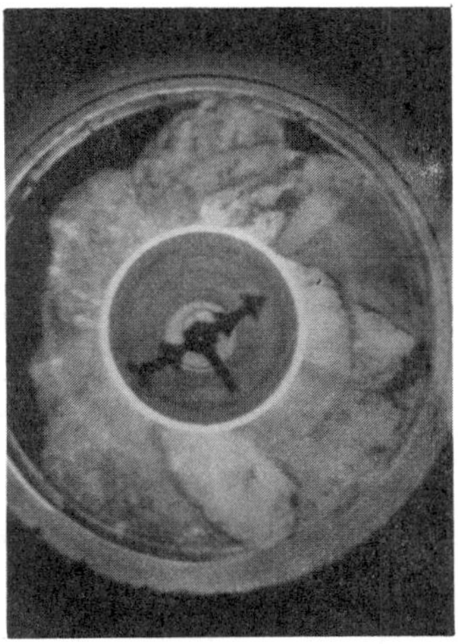

Fig. 19 —Typical borehole photograph showing structure and open fracture.

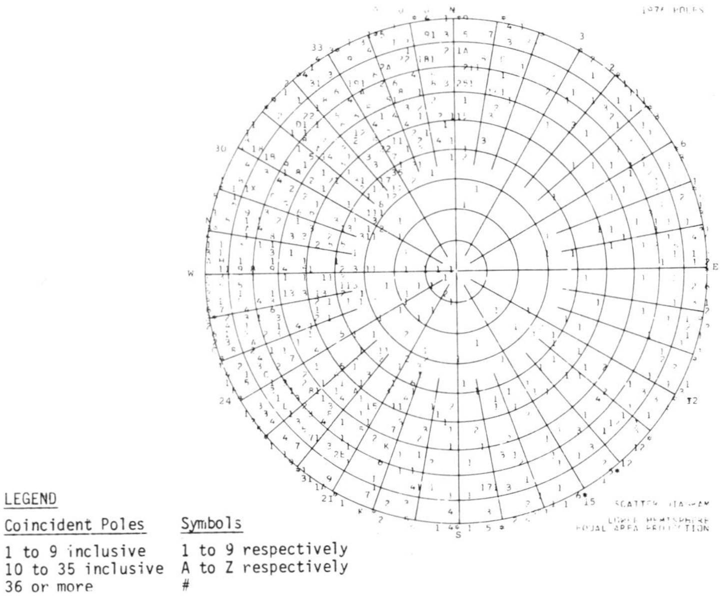

Fig. 20 —Computer plotted polar diagram of 1976 joint measurements.

Samples of rock joints and gouge materials obtained from the boreholes and adits were tested in direct shear in the laboratory to establish the range of shear strength parameters. Within the range of normal stress applied, the angle of friction of the joints ranged from 23° to 36°. In the gouge, the angle of friction averaged 27°. The clay size content of the gouge ranged up to 7%. Typical results of tests on joint surfaces are shown on Fig. 21. Fig. 22 shows a typical shear test in the low stress range for a clayey gouge. Unconfined compression tests were also performed on intact weathered rock. Average strength values were 340 p.s.i.

Bedrock permeability was measured by performing water packer permeability tests in the same boreholes used for the borehole photography. The packers were spaced at 20 ft. intervals during each test. The bedrock permeability values ranged from 6×10^{-5} to 3×10^{-7} cm./sec. In order to monitor groundwater conditions at a later date, piezometers were installed in the boreholes following water packer testing.

A preliminary pit design was developed from this information. Analysis of the structural orientation data was performed to determine potential failure geometry. Relative stability analyses were performed using test data for discontinuities where potential failure zones coincided with a structural discontinuity. Where the potential failure zone was estimated to pass through intact rock, a shear strength value of 300 p.s.i. for weathered rock was assumed. To take into account the influence of groundwater, sensitivity studies were performed to determine the effect of varying piezometric levels on the safety factor. Computer analyses, using the finite element method (7) were used to study seepage conditions, the influence of various types of drainage and the influence of seepage and cleft water pressures on stability. The analyses indicated that controlling the phreatic surface to a depth of about 150 ft. below the slope face would allow pit slope angles to be steepened 3° to 10° in various locations of the pit.

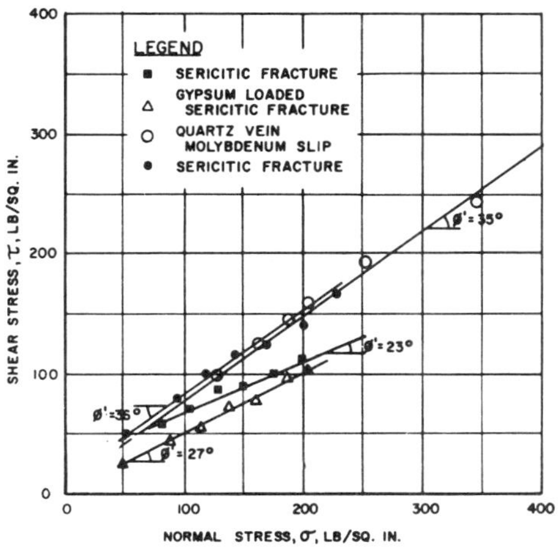

Fig. 21 —Direct shear test results on rock joints (peak values).

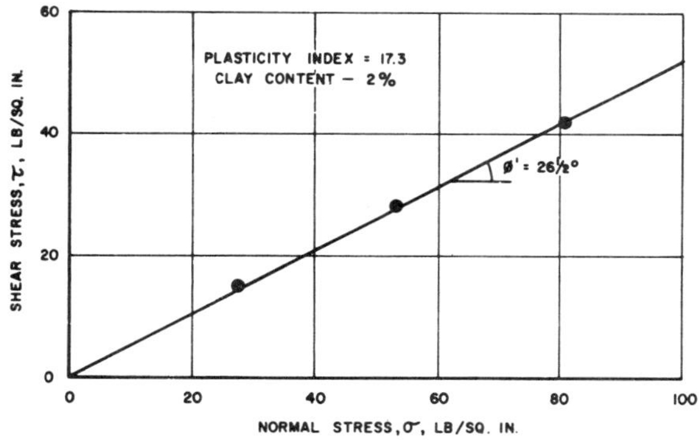

Fig. 22 —Direct shear test results on gouge (peak values).

The type, frequency, and orientation of the structural discontinuities were significantly different in three areas of the pit. The preliminary slope angles listed below were computed for the three zones based on complete groundwater control and the use of a blasting program near the final slope that limits seismic acceleration forces. The use of cushion blasting and angle drilling at the final face was recommended.

Preliminary Slope Angles (Degrees)

Zone	0-500 ft.	500-1250 ft.	1250-2500 ft.
1	50	45	40
2	55	48	44
3	42	38	36

Several simplifying assumptions were made in analyzing the data and in developing the preliminary pit design. Therefore a continuing field observation program must be maintained as the pit is developed to review the final pit design.

The following recommendations were made:

1. Continue to map geologic structure.
2. Install piezometers 200 ft. deep from every third bench at 1,000 ft. spacing and read these weekly.
3. Excavate occasional steep slopes at the working faces three benches high, steepen these until failure occurs and using the 'back analysis' technique compute the average shear parameters. Use these values for continued final slope angle review.
4. Experiment with final pit wall blasting techniques to develop a blasting program that minimizes acceleration forces due to blasting.
5. Implement a subsurface slope drainage program as excess cleft water pressures develop.
6. Develop and maintain a program to monitor slope movements so that early warning of any major movement is obtained.

REFERENCES

1. Dobry, R. and Alvarez, L., "Seismic Failures of Chilean Tailings Dams," J. Soil Mech. and Foundation Div., ASCE, Vol. 93, No. SM6., 1967, pp. 237-260.
2. Brawner, C.O., and Huculak, N.A., "The Use of Horizontal Drains in Landslide Stabilization," Forty Second Convention, Canadian Good Roads Association, Banff, 1961, pp. 383-400.
3. Brawner, C.O., "The Influence and Control of Groundwater in Open Pit Mining," Fifth Canadian Symposium on Rock Mechanics, University of Toronto, 1968.
4. Horn, H.M. and Deere, D.U., "Frictional Characteristics of Minerals," Géotechnique, Vol. 12, No. 4, 1962, pp. 319-335.
5. Jenike, A.W., and Yen, B.C., "Slope Stability in Axial Symmetry," Proceedings of the Fifth Symposium of Rock Mechanics, Pergamon Press, Oxford, 1963.
6. Brawner, C.O., "The Three Major Problems in Rock Slope Stability in Canada," Second International Conference on Surface Mining, Minneapolis, 1968.
7. Zienkiewicz, O.C., and Cheung, Y.K., "The Finite Element Method in Structural and Continuum Mechanics," McGraw Hill, London, 1967.

Chapter 13

PANEL DISCUSSION

Panelists:

Dr. K. Barron, Dept. of Mines, Canada
Mr. C. O. Brawner, Golder, Brawner & Associates, Vancouver
Dr. L. Casagrande, Harvard University, Cambridge, U.S.A.
Dr. H. Q. Golder, H. Q. Golder & Associates, Toronto
Dr. E. Hoek, Royal School of Mines, London
Mr. B. Kennedy, Anaconda Company, U.S.A.
Mr. D. L. Pentz, Golder, Brawner & Associates, Vancouver

H. L. HAMMERSLEY, B. C. Dept. of Mines.

I would like to ask Mr. Brawner if the direction that he measured in the boreholes with the camera was referenced to magnetic north?

MR. BRAWNER.

Yes, it was. However, there is one problem that you may encounter if the ore body is magnetic; the borehole compass will give erroneous results. Three procedures are available to overcome this. The first one involves the use of a gyro compass instead of a magnetic compass. The second technique developed by Christiansen in Salt Lake City, is to orient the core with the use of scribing knives. The third method is to remove the compass needle and place a steel ball in the compass case. The hole must be drilled at a slight angle off the vertical. The location of the steel ball, which will roll to the underside of the hole is photographed. A borehole directional survey is then made, from which the bearing of the steel ball relative to the axis of the compass can be determined. The joint orientation is then related to this bearing at all depths.

J. ROBINSON, B. C. Dept. of Mines.

It was stated this morning that there have been some recent developments that give us the ability to drill horizontal drainage holes and install drainage pipes more cheaply than in the past. Would Mr. Brawner tell us some of the new developments and methods of installing these horizontal drain holes?

MR. BRAWNER.

There are two zones in which you may be concerned about installing drains; the first is in the overburden soil, the second is in the rock. Each area requires a different installation method. If the rock is fairly massive and the holes will remain open, the simplest procedure is to use standard air track equipment or a percussion drill. At one of the mines in the Highland Valley, drains up to 300 ft. long have been installed with a percussion drill at a rate of one drain per day. The costs were in the order of $2.30 and $2.40 per foot. If the rock is quite soft, such as in a serpentine, it will be necessary to install a casing in the hole. The least expensive casing that will not deteriorate within the life required should be used. Plastic pipe or black iron pipe are commonly used. In soft rock, it may be necessary to drill with a casing to prevent the collapse of the hole before the drain pipe is installed.

In overburden, standard diamond drilling equipment has been used but has been modified to install the drains. The technique normally used is to drill with an H casing and a hollow tungsten carbide bit. The hole is drilled to the required depth, pipe is installed inside the casing, and then the casing is removed.

Occasionally you run into awkward problems. For example, if you have running sand, you may find that the friction from the sand is so great that you cannot withdraw the pipe. In loose silt, the return wash water may erode excessive amounts of silt from around the pipe. In plastic clay soils, the clay may ball up and plug the pipe. In this case, air has been used successfully as a drilling fluid. In my experience, serious difficulty with installation techniques occurs in about one case in five.

MR. L. CROUCH, Dept. of Mining Engineering, University of British Columbia.

Does a great variation in climate influence the approach to stability studies?

MR. BRAWNER.

Yes, it does. In general, the valuation should take into account the most unfavorable climatic conditions expected at the site. In the areas where snow falls in winter, the spring thaw period is normally associated with high groundwater levels. In northern climates, the cold temperatures cause the slope to freeze so that the water, which normally seeps from the slope, can no longer get out as easily. This, in a sense creates a dam behind which high pore water pressures can develop, thereby reducing stability.

The majority of slides at open pit mines, that I have investigated in Canada, have developed during the winter months. In one case, in the Highland Valley of B.C., a slide developed on a day when the temperature dropped to -57°F. in the pit.

Tension cracks are often used as an indicator of instability. However, during wintertime, snow may hide cracks that develop.

High pore water pressures also frequently exist in areas where very heavy rainfall occurs.

DR. HOEK.

I would like to add one comment to Mr. Brawner's discussion, and this is to emphasize the importance of tension cracks being filled with water. I've recently had experience in a limestone quarry in England, where the dip of the rock structure was relatively flat, about 18°. Looking at the structural conditions, there appeared to be no reason why a very violent failure had occurred. But, in fact, the failure had occurred after a period of about 3 weeks of incessant rain and a heavy rainstorm. A vertical joint had opened up, filled with water, and developed a very high horizontal force, which had pushed a large block out that demolished everything in its path. It must be emphasized that water filling vertical tension cracks can be an extremely serious problem in some slopes.

One of the things I'd like Mr. Brawner to comment on is the necessity of controlling surface water and the filling of tension cracks, and whether, in fact, one can effectively drain the toe of the cracks by horizontal drainage.

MR. BRAWNER.

Anything that you can do to keep water out of the slide area is beneficial. Horizontal drains can be very effective in reducing water pressures in tension cracks. It must be recognized that tension cracks normally do not extend more than 1/4 to 1/3 of the depth of the slide, so that for this problem the horizontal drains must be installed near the bottom of the cracks, not at the toe of the slope.

As cracks develop, movement markers should be established on both sides of the cracks and the cracks should be filled to prevent the entrance of water.

DENIS BRATTON, Hilton Mines Ltd., Quebec.

I would like Dr. Casagrande to comment on the methods and extent of compaction required for tailings dams, and the effect of compaction on permeability.

DR. CASAGRANDE.

There are many types of equipment that can be used for compaction. For granular materials, we can use a heavy crawler-type tractor or vibratory rollers. The degree of compaction is very important. On large projects, I feel we should make large scale field investigations to determine the most economical type of equipment suitable for a specific type of material to be compacted. In addition, we want to know the most economical thickness of layer that can be compacted for the specific type of equipment selected.

For small projects, we do not require expensive preliminary investigations. If granular soils are to be compacted, we could simply select a D7 caterpillar tractor for compaction and perform some field density tests to measure the density we are getting. Relative density tests are recommended for this purpose. There is some difference of opinion regarding the degree of density necessary. In earthquake areas, some engineers, based on long standing experience, consider 60% relative density necessary so we do not have to worry about liquefaction. On the other hand, the U.S. Atomic Energy Commission insists on 80% relative density. As far as compaction of mine tailings dams is concerned, if we can obtain a compaction effort of 60% relative density, I think that this is sufficient.

BOB DOWDELL, Cominco Limited.

I have two questions for Mr. Brawner. From the borehole photography and the shear tests that were conducted at the last site mentioned in his paper, I would like to ask how the pit slope angles were developed from such measurements and what degree of confidence can be assigned to the design slope angle to assure the operators that their pit slope will not fail?

MR. BRAWNER.

The most important factor in evaluating the pit slope angle, for the last project that I discussed, was an evaluation of the existence and orientation of the joint and fault systems of the rock structure. Statistical analyses of over 10,000 structural measurements were made to determine in what areas, if in any, the orientation was adversely oriented in relation to the pit face. In all areas of the pit slope, the majority of the discontinuities was quite steeply inclined. For most of the pit circumference, there was one system of discontinuities located relative to the pit face so that it could exist as the back portion of a potential slide wedge or block. Shear tests on joints were performed in the laboratory of rock on rock, gouge on rock, and in the gouge. The structural measurements did not reveal any significant flat lying joint systems. Therefore, for a failure to occur, the movement would have to occur through some intact rock. Unconfined compression tests were performed on numerous rock samples of the intact rock taken at depth. Some samples exhibited noticeable alteration or weathering. The values of friction angle and unconfined shear strength selected for design were the 10% values (90% of all test data exceeded the values selected). The slopes were also analyzed for failure by toppling and buckling.

Regarding the second question, the first responsibility of the designer is to consider how serious a failure could be. As the degree of seriousness increases, the analysis and design must become more conservative. With the present state of our knowledge, an accurate factor of safety or degree of confidence cannot be computed. In order to provide a reasonable degree of confidence that a designed slope will not fail, all data used in the analysis should be selected conservatively. The economics of various slope angles should also be studied in conjunction with this design. This should consider slopes 5° to 10° flatter than recommended. If these flatter angles result in an economic mine, no further study should be necessary. If they do not, a more detailed study may be warranted. In any event, a program to monitor slope movements should be incorporated in any mine development program so that stability can be continuously assessed.

As indicated in the introduction, some slides should occur if we accept the premise that controlled stability is good mine management.

MR. W. A. WAHLER, Palo Alto, California.

Dr. Casagrande, have you found it necessary to include positive drainage within a tailings-type embankment?

DR. CASAGRANDE.

I have seen a number of tailings dams where particular attention was paid to drainage. Drainage can only help. However we should always ask, is it necessary? If we find it is not necessary, then we can save the expense. We must remember that in spite of the best drainage, the surface of the dam may be sensitive to severe erosion and frost action with all the consequences, unless we protect the surface of the slopes. This protection must be capable of retaining the finer materials. This would comprise a filter zone and then, perhaps, very coarse granular material such as rock or shale.

MR. W. A. WAHLER.

About 30 years ago there was an earthquake in Chile that caused a failure of a tailings dam which killed about 30 people. As a result, one mining company had rudimentary drains installed in their tailings embankment. We had the opportunity, after the dam was 150 ft. high, to investigate the stability as compared to undrained embankments. We found that the drainage was extremely effective and also very inexpensive. New and enlarged embankments were subsequently required. These were designed with quite extensive internal drainage systems, a procedure that was less expensive than some of the other preventative measures that one could use in a highly seismic area.

DR. CASAGRANDE.

I think that this is a very fortunate experience. The only uneasiness that I have about drainage systems is, that you can do a good job, and you can do a poor job. I have seen one out of ten cases where a drainage system worked well. It was well constructed. But how about the other nine? Many drainage systems do not function properly. They become clogged in no time and frequently are not designed so that you can clean them out. The pipes, if not constructed properly, can break, which may lead to piping. However, don't get me wrong, there is nothing wrong with drainage, provided it is done well. If you find that you can afford to install it, by all means do so.

MR. W. A. WAHLER.

To take care of the breaking problem, you can use plastic pipes with slip joints. It should also be borne in mind that in addition to proper design, inspection, and control, that follow up surveillance is necessary. This requires an interchange between the operating people and the design professionals. The finest design is worthless if the construction and inspection are not adequate.

DR. CASAGRANDE.

I fully agree with you.

KEITH ROBINSON, Golder, Brawner and Associates, Vancouver

If the factor of safety of a tailings dam is believed to be inadequate, there are instances where it may be possible to locate a second dam to catch tailings that may be released in the event of an upstream failure.

D. RUSSELL, Manitoba Dept. of Mines.

Is there any way that a person can determine if a soil is sensitive by the appearance of it in the field? One thing that has been brought out in these sessions is that most people do not recognize that they have a problem until after it's happened. Then they call in an expert.

DR. CASAGRANDE.

Mine tailings, if saturated, are sensitive unless they are compacted, and the slimes or the finest portions of mine tailings are very difficult to compact.

As far as soils such as clays or clay-like materials are concerned, if they are sensitive, there is an appreciable difference between the undisturbed strengths of the material and the remolded strength. Another measure of sensitivity can be made by the simple water content test. If you determine the liquid limit of the clay, and find that the water content is appreciably above the liquid limit, then the soil is sensitive. Certain silts are also sensitive to disturbance. As a rule, you should be very suspicious if you find a clay that has strength when you break it and becomes liquid in your hands when you remold it.

In Canada, there are many locations where sensitive clays exist, far more than in the States.

DAVID B. CAMPBELL, Golder, Brawner and Associates, Vancouver

In the back analysis of a failure in a rock slope, assuming that you have measured and know the water pressures on the failure surface, how do you separate the proportion of the strength that may be attributed to cohesion and the portion that may be attributed to friction?

DR. HOEK.

You have to assume one of the parameters, either friction or cohesion. You have no other alternative. In most cases, we can obtain a much closer approximation of the angle of friction than we can of the cohesion. What I normally do, based on my knowledge of rock properties, is to guess what the friction angle might be, or if the joint has infilling, what the properties of that infilling might be, and then back analyze to compute the cohesion. A sensitivity analysis should then be made to see what effect changes in the friction angle have on the cohesion. Generally, the friction angle is a much more definable quantity than the cohesion. However, you have to assume one of the parameters.

MR. L. CROUCH, Dept. of Mining Engineering, University of B.C.

The question of safety factor has come up several times. Could the panel explore the question of safety factors as applied to slopes.

DR. HOEK.

The factor of safety in slopes is defined as the ratio of available strength or the strength that the rock actually has, to the strength that is required to prevent failure. One thing we have to recognize is that the factor of safety is not a magic number in itself. You can't simply say, "If I have a factor of safety of 1.3 then the slope will be safe indefinitely". There is a range of factors of safety and of meanings that we can attach to this. Professor Casagrande gave the example that in dams you are concerned with long term stability of a structure that could endanger life and property. Therefore, one normally uses a factor of safety of 1.5. In rock slope design, I have found in my calculations that a factor of 1.3 is one which results in some individual bench failures that the mine operator finds he can deal with. Dr. Golder mentioned the case where you have had a failure and can do a precise back analysis. You know the material properties accurately, and are therefore justified in using a lower factor of safety; 1.2 was the case he quoted. One has to recognize that the factor of safety is, to a certain extent, a factor of ignorance.

I started my life as a mechanical engineer. In many mechanical engineering structures we can calculate very precisely what the stress conditions are. We know the strength characteristics of the steel or aluminum very accurately, so we can work very close to the limit of either deformation or failure. In rock materials, we can't do that. We don't know the properties of the rock well enough, so we have to build in a conservative estimate all along the line. As Mr. Brawner has pointed out, you have to assess each stage of the calculation and evaluate how much conservatism is necessary. In certain circumstances, a factor of safety of 1.1 can be justified, whereas in other instances a factor of safety of perhaps 2 may be necessary. You normally put in an average value that brackets the range in which you are hoping to work, and you hope that it's always greater than 1.0.

DR. GOLDER.

I want to make two points. The first is that, in the case I quoted this morning, where we used a factor of safety of 1.2, it wasn't because we liked it; it was because it was the highest factor of safety we could get. There wasn't room to place any more soil at the toe of the slope. The fact that we had a failure did mean that we could determine the shear strength with sufficient accuracy to know that we were not going to get another failure. My second point is that you have load factors that you apply to loads, factors of safety that you apply to materials, and factors of ignorance that you apply to engineers.

MR. B. KENNEDY.

One of the things that worries me, is that the factor of safety that I calculate probably differs from yours, because I have used a different interpretation of my input data. Therefore, there is a probability problem. When evaluating any variable, should we design on the basis of the mean value (which in terms of strength, 50% failure) or should we design, say, at two standard deviations below this, which will give a 98% probability of safety? With several parameters we have a multivariable problem. Until we define these limits, it is always going to be difficult to assess the overall factor of safety.

MR. PENTZ.

I would like to add one further point to the comments that have already been made. If we assume a limit equilibrium type of analysis, then we also assume a particular shape or type of failure surface so that when we carry out a computation of the factor of safety, not only are we inherently assuming certain strength parameters to actually be available, but we are also assuming that a particular failure surface will be there.

In the case of a circular arc analysis, we can examine many different circular arcs so that we can then find the circular arc with the minimum factor of safety for a particular strength parameter and groundwater condition.

DR. GOLDER.

If you do an analysis, you cannot start at the beginning and say, I am going to need a factor of safety of 1.5, because the answer you get depends not only on the failure parameters that you feed into your calculation but on the method of analysis that you use. Let us say we are going to analyze by what is called 'the method of slices'. There are at least three different 'methods of slices' in common use. You have the simple Swedish method, where you neglect the forces between the slices; you have Bishop's variation; and you have recent work by Morgenstern. The difference in factor of safety between these can be as much as 20%. I have seen analyses done for an earth dam where everyone said they wanted a factor of safety of 1.5. If you want 1.5, all you've got to do is use the 'correct' method of analysis. The real factor of safety you don't know. The more accurate your method, the closer you are to the truth and, therefore, you can use a lower factor of safety.

Another important point is that if you have a material that has two parts to its strength, i.e. cohesion and friction, then in your calculations you do not leave this figure F lying about on one side of the equation. You must divide the cohesion by F, and the coefficient of friction or tan ϕ by F.

MR. B. KENNEDY, Anaconda.

I would like to make one further point. In the paper that Dick Stewart and I presented, we included a chart of probability of failure against slope angle for risk analysis in design. That chart is a very simple example. If this chart was drawn up with all the factors considered, you wouldn't have just one line. You would have a number of lines, depending on what analysis technique you use and what factors you select to use for your various input parameters. In the design phase for mine feasibility studies, this chart will be used in a risk analysis with a large number of other factors including ore grades, cost of mining, price of the metal, etc., which are plotted on a similar probability of occurrence chart. These are evaluated two at a time, and worked one against the other until a final risk is developed for the property. Therefore, rather than giving the mine engineer one factor of safety, it would be more useful to give him a range of probabilities of failure.

MR. DERMOT ROSS-BROWN, Imperial College, London.

I would like to turn the tables on the panel, and ask them to imagine that they themselves are mine managers of a medium-sized mine in Canada. By medium size, I mean a mine producing about 5-10 million tons of ore and waste a year, and employing about 300 people. The upper walls of the open pit consist of 100 ft. of overburden, standing at about 35°, and below this is a fairly competent rock, standing at an overall angle of about 45°. At the moment you are only down about 250 ft., but the planned ultimate depth is more like 1500 ft. To date, little or no thought has been given to slope stability studies or the collection of any relevant data. While you appreciate the need for such studies, unfortunately, your Board of Directors doesn't share your view at the moment. It is unlikely that you will get any budget allocation from them for at least 12 months, but you can probably spare $2000 from another account. As far as manpower is concerned, you could probably spare a geologist and a mining engineer for up to 20% of their time to work on such studies. Although it is a hypothetical question, I don't think it is untypical of the situation in a lot of North American mines at the moment, and I would ask the panel to suggest what studies they would initiate and develop during the next 12 months, given these conditions.

A second question, I would like to ask is what they would do if they had an annual budget allocation of $40,000?

MR. BRAWNER.

Regarding allocation of $2000 for 12 months, my first thoughts would be to organize a structural mapping program with the geologist. This would comprise representative mapping of faults and shears exposed in the pit, presentation of this data on punch cards and computer printout of polar diagram plots of the data. Secondly, I would utilize the drilling equipment available at the mine to drill about 10 holes around the working areas of the pit and install piezometers to measure water pressures in the slope. Thirdly, I would arrange with the blasting foreman to perform some experiments with pre-shear and cushion blasting, in selected areas around the pit, to develop blasting techniques that would minimize damage to the final slopes. If someone else had some free time, I would request a surface hydrologic study to assess groundwater conditions around the pit and develop a program to control surface water.

DR. BARRON.

I concur with Mr. Brawner's approach. In addition, I have two further suggestions. One would be to send the mine manager on a trip to visit several deep open pits where rock slope problems exist and where stability studies are being performed. In addition, I would put the geologist to work reorganizing some of the information that exists. This would include a review of the initial drill logs that were obtained for the feasibility study for the pit, such as relogging the data and calculating RQD values.

Regarding the $40,000 question, this amount is slightly less than what one might assume as a budget for a 2-man slope stability study. The U.S. Bureau of Mines, 5 years ago, were using a figure of $25,000 per year per man for the men and equipment on a slope stability study. I would invest in some first class survey equipment, perform detailed geological studies, drill or extend boreholes to a depth of 1500 ft., and perform a hydrologic study. Analysis would involve considerable computer time.

MR. PENTZ.

I would like to make some additional comments. Early in the program I would install a simple displacement monitoring system, so that if any failures did start, the displacement characteristics of those failures could be picked up and a back analysis of the stability could be undertaken. In addition, I would develop a program to evaluate any small bench failures that occur. If any exploration or drainage adits exist in the area of the pit, these should be mapped. If any adits are proposed, holes should be drilled and piezometers installed in advance of the adit. The piezometer records, taken as the tunnel advances, can be used to compute the permeability and potential influence on drainage.

DR. HOEK.

One of the problems in trying to suggest solutions to Mr. Ross-Brown's hypothetical question is that it presupposes that there are people around who know how to handle the data you can collect, whether it be for $2000 or $40,000. One of the real problems that we face in this field is that if I asked you to recommend a particular text, paper, or conference proceedings that outlined all the things that you could do, there just isn't one. A major problem in education exists. There are very few people who know much about this subject. The mining schools have not taught it adequately, and it may be necessary to rob the civil engineering schools. In my opinion, one of the most important things is to try and ensure that you have somebody available who knows what's going on. I believe that it is absolutely essential that there be some in-house capability in a mine. I think it is inefficient to call in a consultant and simply present him with your problem and say, "There it is. Solve it". He is at a disadvantage unless the mine has someone to provide a background of the geologic and hydrologic information he will need for the particular property. All of us have simply assumed that you start off from this data and that you have somebody who knows how to handle the problem. I don't think this is necessarily a valid assumption at many mines. This is not knocking any person in particular but it does suggest that knowledge hasn't been transmitted from the experts to the mine engineer. There is a lack of communication and it is time that this was overcome.

DR. GOLDER.

I am older than all these men, and I would treat this problem quite differently. It's a very good question. With the $2000 I'd throw one hell of a party for the Board of Directors and when they were all drunk I'd get them to authorize the expenditure of $40,000.

DR. B. SEEGMILLER, Anaconda.

I have a question concerning water. On Monday, Mr. Brawner stated that contrary to common belief, water does not generally act as a lubricant in rock. I would like to hear the panel comment on that statement.

MR. PENTZ.

I would like to emphasize that this is a very common misconception. The term lubrication has quite different meanings in different branches of engineering. Dr. Morgenstern showed very clearly that in lubricating a human joint, lubrication is essentially an effective stress principle. When we talk about water not lubricating a slide, we mean that water does not chemically alter the material and thereby reduce the basic friction coefficient.

The influence of water is to reduce the normal stress on the failure plane. However, where you have sensitive gouge materials, or where considerable movement is involved, the friction angle may reduce from a peak value to a residual value as a result of shock or excess strain. What is meant when we say that water does not lubricate a slide, is that for most rock surfaces, water causes a reduction in effective stress rather than a reduction in the friction angle.

DR. HOEK.

It is possible, in highly weathered rock, that a change in moisture content may produce a decrease of friction coefficient.

DR. GOLDER.

Laboratory tests have shown quite clearly that if quartz is wet, the coefficient of friction is the same, to within 1°, as it is when it's dry, so it's not a lubrication effect that reduces the strength. There is quite clear experimental evidence that shows that except for some serpentine and schists, water does not generally reduce the angle of friction in rock. When you have an increase in water pressure, the total pressure remains the same, but the effective pressure is reduced and that is what controls the shear strength of a rock or a soil.

DR. R. CALL, University of Arizona.

There is one factor in slope stability that has not been touched on in this meeting. The factor of regional or tectonic stress. I'd like to hear the panel comment on its effect on stability and the method of measuring and analyzing it.

DR. HOEK.

This is a problem that concerned me very much when I started in slope stability, because I came to it from the experience of hard rock underground mining in South Africa. The first thing I said is that you've got to measure the stress and do a stress analysis. We have looked at this problem theoretically and with models. It appears to have a large influence on the overall displacement pattern around the mine, which is important. This displacement might influence the permeability by local opening of joints due to relaxation of the stress. However, when we look at the actual failure surface itself, and the wedge of rock within close proximity of the slope face, our experimental work has shown that the horizontal stresses are almost totally relieved by relaxation resulting from the slight opening of discontinuities. Therefore, the calculation, on the basis of limit equilibrium in a static situation, assumes only vertical gravitational forces and appears to be the one that most closely approximates the actual failures. This is why, in most of these analyses, we ignore tectonic stress fields. I should add though, that where you are concerned with the effect of the pit shape on the slope stability, that you could have differences depending on the direction of the tectonic stress field which would superimpose on the arching effect of the pit shape.

MR. BRAWNER.

I agree with what Dr. Hoek has said. There is one instance, however, where we can run into difficulties. That is where the rock has essentially a horizontal structure and the terrain is flat. This condition exists in the limestone quarries around Montreal. As a quarry is widened and deepened, usually with near vertical slopes, buckling failures frequently occur in the base of the pit. This condition is partly due to tectonic stress.

In the porphyry type copper mines that occur in British Columbia, there is a high degree of faulting and jointing. This suggests that tectonic stress has, to a large degree, been relieved by movement. In this type of deposit, we do not consider that tectonic stress has an important influence on stability. A quasi tectonic stress can occur as a result of the great imbalance of weight, however, when an open pit is located in a valley between high mountains.

MR. PENTZ.

I would like to make one point regarding the measurements of stress. Dr. Barron pointed out, in his paper on instrumentation, that there is a lot of controversy as to whether there are instruments that can measure stress with reliability. I agree with this, and if it is true in unfractured rock, then it is more true in fractured rock. I suspect that the basic problem is that a lot of people have postulated that tectonic stresses do exist. Several authors have said that it has been measured, but I know of several court cases where, in fact, this is in doubt. While we suggest from the basic mechanics of the earth's crust that a tectonic stress exists, there is doubt that it has actually been measured close to an open pit surface.

DR. BARRON.

I believe that tectonic stresses have been determined on the basis of measurements, but I can't say how accurately. These have been mainly in underground situations, more often in deep underground operations. In open pit mining, I would not be inclined to spend my $2000 or $40,000 on this type of investigation. On the other hand, I think it is a viable area for research work. It is something we should not ignore completely. Perhaps the realistic use of the finite element technique will help evaluate this problem.

DR. CALL.

I concur with Dr. Barron's comments. We've been performing some strain relief studies in a porphyry copper deposit that we've been investigating for input in a finite element analysis, and we have concluded that residual stress exists. Basically, I agree with Dr. Hoek, that for the analysis of the majority of pit slopes, that it is not important.

GORDON HILCHEY, Sumitomo Metal, Vancouver.

I was wondering if the panel would care to give an outline of that data that they would like to have available, as consultants, if they were to appear on a property, in the feasibility study stage; data that was obtained early in the exploration stage.

MR. BRAWNER.

The most important initial step is to review all available geological data. In addition, selected boreholes should be saved so that they could be used at a later date, if necessary, for a borehole camera or borehole television survey, for water packer permeability testing, or for the installation of piezometers. This will require a casing to be left the full depth of the overburden. When your core is recovered from the hole, and prior to splitting for assay, the core should be logged structurally and photographed in color. Four or five core boxes can be photographed together.

I strongly urge that triple wall core barrels be used for all drilling where core recovery falls below 95%.

DR. HOEK.

One of the areas in which we in rock mechanics lag far behind soil mechanics is that, in general, we don't make adequate use of the exploration program. In the evaluation of a site for a civil engineering project, there are a great number of simple identification indexes that can be applied at an early stage of the project that can help the engineer assess the conditions he has and determine if more detailed testing is required. In general, we don't do this in rock. We tend to drill for mineral evaluation only and forget about stability until later when we have a potential pit on our hands. Recognizing this problem, and recognizing the shortage of an adequate number of competent people, at an early enough stage in the program, we at Imperial College have been looking at the possibility of very simple index tests. A typical example is a point load test, which is actually a core splitter with a load measuring device. It is desirable that the geologist, in the early stage of an investigation, do a few selective simple index tests, which have some meaningful correlation with rock properties and which could provide assistance at a later stage to define precisely what sort of problem one wants to look at in more detail.

DR. BARRON.

In coal mining exploration, coring is very often not practised. Rather, cuttings are examined and tested. If we must, at a later stage, start thinking about the safety of pit slopes, we must then do some structural drilling. It might well be better to combine the two in the first program. This will provide more information at an early stage with less overall expense.

MR. PENTZ.

I would like to note that air photography can also provide important geologic information. Major fault lines and structural lineations often can be observed and used in establishing the most feasible drilling and site programs.

Chapter 14

CONFERENCE SUMMARY

by

Evert Hoek

Professor of Rock Mechanics, Royal School of Mines, London

Summarising a conference is never an easy task, but I will try to keep my comments brief and to the point. There are a number of lessons which we have learned from our deliberations together during these past three days, and I will try to give you my impressions of these lessons and of their practical implications in open pit mining.

The first thing that has become increasingly obvious during these discussions is that it is no longer acceptable to offer an 'Act of God' as an excuse for a major slope or tailings dam failure. Disasters such as that at Aberfan in the United Kingdom have focussed the world's attention on the activities of the mining industry and the report of the commission set up to investigate the Aberfan disaster clearly showed that adequate knowledge was available and that failure to apply this knowledge did not qualify as an 'Act of God'. The courts and the world's press are certain to impose increasingly harsher penalties and criticism on those held responsible for major failures, particularly of tailings dams, in the future.

One of the major problems in applying available knowledge to slope problems is that attempts to apply this knowledge are frequently left too late. To provide an engineering solution in the case of a slope in an advanced state of failure is an almost impossible task, and it is essential that those responsible for the stability of a slope must learn to recognise signs of distress at an early enough stage for effective remedial measures to be taken. As our ability to solve slope problems improves, the time at which slope stability problems are considered in a mining project must move forward to a stage at which these problems can be anticipated and dealt with before they arise. Conferences such as this are important in that they enable us to exchange ideas and advance our knowledge to a level at which such predictive engineering will be possible. It has been encouraging to hear of mining projects such as the Bougainville Copper operation where slope stability studies were carried out as part of the overall evaluation programme and it is hoped that this will become a more common procedure in the future.

Providing a safe slope in an opencast mine obviously costs money, either in terms of the reduced ore to waste ratio given by flat slopes or because of the cost of stabilization techniques such as drainage or the use of rock anchors. An optimum slope design will provide maximum safety at minimum cost but achieving this optimum is not an easy task. Any one of us could design a slope which would be stable to the end of time, but it is probable that that slope would be so flat as to be totally uneconomic in most open pit mining situations. On the other hand, if the overall economics of a mining project are so critical as to require slope angles which the engineer responsible for these slopes considers to be unsafe, it is his duty to make this clear in the strongest possible terms and so prevent the development of a potentially disasterous situation.

Stewart and Kennedy, in their paper, have given us some idea of the possible costs of slope failures and it would be as well to keep these costs in mind when, in order to maximise profits, one pushes the slope angle as steep as one dares.

An important aspect of the economics of slope stability is that provision must be made in the initial project budget work for any investigations or special services which may be required. It is certainly more efficient in the long run to spend a little more money on the early mineral exploration drilling in order that information required for the analysis of slope stability should be collected at this stage rather than in a separate drilling programme carried out later. How much money, you may ask, should be allocated for this work. This is a very difficult question on which to give a general answer since so much will depend upon local conditions. However, experience tends to suggest that approximately 1% of the total capital investment is the order of investment necessary to ensure that all reasonable steps are taken to collect information and to provide a sound slope design.

It has been said that each problem in rock engineering is controlled by a unique combination of rock structure, rock strength, groundwater conditions and other factors of influence and that, therefore, one is not entitled to extrapolate from one situation to another. While it is true that rock is not a man made material with predictable properties, it is, in my opinion, unnecessarily pessimistic to suggest that the lessons learned on one problem cannot be applied to another problem. Certainly, one cannot simply decide upon a slope angle for Mine A on the basis that Mine B, which may be a thousand miles away, has never had slope problems and its slope angles must therefore be safe. On the other hand, it may well be possible to examine a number of mines in the same rock type as that which exists in Mine A and, from a careful analysis of points of similarity and points of difference, draw very useful practical conclusions.

The justification for much of the theoretical work on rock slope stability which is currently being carried out in research organisations and universities around the world is that it should provide a mathematical framework upon which we can hang most of our practical observations. If the mechanics of rock slope stability can be adequately represented in mathematical terms, then it should be possible to link the practical experience gained in a range of different rock environments. And this practical experience is one of the most valuable assets which we possess. The most important function which we research workers can perform is to provide a vehicle whereby you practical mining engineers can communicate your experiences to each other.

Another aspect of practical experience is the analysis of existing slope failures such as provided by C. O. Brawner. No theoretical model or experimental laboratory study can ever reproduce all the conditions which occur in a slope failure and yet most research workers tend to confine their activities to the laboratory — ignoring the vast supply of data available in the field in the form of actual slope failures. Admittedly, it is necessary to spend long enough in the laboratory to sort out ones ideas on what to look for in the field, but, in my opinion, our ultimate ability to design slopes will depend, to a large extent, upon our ability to make full use of that information which is already available in the field.

An extremely important point was raised by Professor Casagrande in his discussion on a set of design charts developed for checking the stability of tailings dams in South Africa. He pointed out that, while these charts were undoubtedly most useful under the particular conditions for which they were developed, they should not be used as a general guide for tailings dam design. I have myself, recently produced a set of design charts for slope stability but I have taken care to list the assumptions made in deriving these charts and the limitations in their application and the charts should not be used outside these limits. It is all too easy to take a set of design charts or design formulae and to apply these to a particular problem. An answer will be obtained, but unless the design method has been used for the conditions for which it was derived, this answer will be meaningless.

It is up to us, the academics and research workers, in this field to define, in very precise and clear terms, the conditions and limitations of any design methods which we propose. It is up to you, the practical engineers concerned with these slope problems, to read these conditions and limitations and to use the slope design methods only within these limitations.

Turning now to what I consider to be the greatest single difficulty in rock slope engineering — the lack of adequate geological information. It has been made abundantly clear during this meeting that rock slope stability is, above all, controlled by structural discontinuities in the rock mass, and that any meaningful analysis of rock slope stability must take the geological conditions of the site into consideration. And yet, how often do we attempt to arrive at a solution to a slope problem on the basis of minimal geological data. It has been said that attempting to solve a rock engineering problem without adequate site investigation is equivalent to asking a surgeon to operate without having any X-rays of a patient available, and I believe that to be a very fair comparison.

It must be admitted that, at present, our techniques for the collection of data on rock structure are inefficient, and that there has been too little communication between engineers and geologists in an effort to define what structural data is necessary for slope stability analysis. However, the development of improved methods of stability analysis will depend upon developments in this area of geological data collection. There is no doubt in my mind that our present computational ability far exceeds our ability to collect meaningfull input data, and I would suggest that there is little point in further development of our theoretical techniques unless there is a parallel development of data collection technique.

The influence of blasting on rock slope stability is a factor which appears to have been largely ignored in the past, but, with increasing size of blasts, is an influence that we can no longer afford to neglect. I have seen a number of slopes around the world which have been so heavily damaged by blasting that any attempt to apply conventional methods of stability analysis to them would produce meaningless results. Heavy blasting may produce results which are ideal from the transportation and crushing point of view, but they tend to leave a severely damaged and probably unstable slope behind.

Fortunately, the use of pre-split blasting to form the final slope before the heavy blasting is carried out tends to limit the damage to the remaining slope and I believe that it is in everyone's interest to examine the use of pre-split blasting in opencast mining. It may be expensive to implement if the actual cost of pre-splitting alone is considered, but I feel that it could prove an economical process if this cost is offset against the improvement in stability and the reduced maintenance costs of the slope.

An allied subject was dealt with by Professor Whitman in his excellent paper on the influence of earthquakes on slope stability. Most of us simply do not consider earthquakes because we do not understand them, but Professor Whitman has shown that it is possible to obtain some estimates on the earthquake risks in a particular area and to increase the factor of safety of the slope to allow for this risk.

Several contributions, including Professor Morgenstern , have dealt with the influence of groundwater on slope stability and of possible drainage measures which can be applied to reduce this influence. Similarly, Dr. Golder has lead our discussion on to what we can do to improve the stability of slopes by grouting and bolting. I will not dwell further on these topics, not because I wish to imply that they are unimportant, but because I believe that all of us appreciate their importance and recognise the need for further work in these areas.

Professor Casagrande gave us the benefit of his considerable experience in tailings dam design and his outstanding lecture must surely be one of the highlights of this meeting. The obvious lessons which his paper carries for all mining engineers do not need further emphasis by me.

Finally, we come to the outstanding example presented by Bruce Kennedy on the prediction of the Chuquicamata slide. It is an encouraging thought that if things go wrong in an opencast mine and all possible steps have been taken, but it is obvious that a massive failure